AF903485

PROGRESS IN STEM CELL TRANSPLANTATION

Edited by **Taner Demirer**

Progress in Stem Cell Transplantation
http://dx.doi.org/10.5772/59336
Edited by Taner Demirer

Contributors
Khalid Al-Anazi, Asma Al-Jasser, Franca Fagioli, Jisook Moon, Snag-Hun Bae, Osman Ilhan, Sinem Civriz Bozdag, Zhiquan Shu, Shelly Heimfeld, Zhongping Huang, Dayong Gao, Jihwan Song, Hyun Sook Kim, Pervin Topcuoglu, Ana Rita Caseiro, Tiago Pereira, Paulo Bártolo, José Domingos, Ana Colette Maurício, Ana Lúcia Luís, Darda Bayraktar

Published by InTech
Janeza Trdine 9, 51000 Rijeka, Croatia

© The Editor(s) and the Author(s) 2015
The moral rights of the editor(s) and the author(s) have been asserted.
All rights to the book as a whole are reserved by InTech. The book as a whole (compilation) cannot be reproduced, distributed or used for commercial or non-commercial purposes without InTech's written permission. Enquiries concerning the use of the book should be directed to InTech's rights and permissions department (permissions@intechopen.com).
Violations are liable to prosecution under the governing Copyright Law.

Individual chapters of this publication are distributed under the terms of the Creative Commons Attribution 3.0 Unported License which permits commercial use, distribution and reproduction of the individual chapters, provided the original author(s) and source publication are appropriately acknowledged. More details and guidelines concerning content reuse and adaptation can be found at http://www.intechopen.com/copyright-policy.html.

Notice
Statements and opinions expressed in the chapters are these of the individual contributors and not necessarily those of the editors or publisher. No responsibility is accepted for the accuracy of information contained in the published chapters. The publisher assumes no responsibility for any damage or injury to persons or property arising out of the use of any materials, instructions, methods or ideas contained in the book.

Publishing Process Manager Iva Simcic
Technical Editor InTech DTP team
Cover InTech Design team

First published December, 2015
Printed in Croatia

Additional hard copies can be obtained from orders@intechopen.com

Progress in Stem Cell Transplantation, Edited by Taner Demirer
p. cm.
ISBN 978-953-51-2227-2

We are IntechOpen, the world's largest scientific publisher of Open Access books.

3,250+
Open access books available

106,000+
International authors and editors

112M+
Downloads

151
Countries delivered to

Our authors are among the
Top 1%
most cited scientists

12.2%
Contributors from top 500 universities

WEB OF SCIENCE™

Selection of our books indexed in the Book Citation Index in Web of Science™ Core Collection (BKCI)

Interested in publishing with us?
Contact book.department@intechopen.com

Numbers displayed above are based on latest data collected.
For more information visit www.intechopen.com

For EU product safety concerns:
IN TECH d.o.o., Prolaz Marije Krucifikse Kozulić 3, 51000 Rijeka, Croatia,
info@intechopen.com or visit our website at intechopen.com.

Meet the editor

Dr. Demirer graduated from Ankara University Medical School in Turkey in 1984 and trained in the USA from 1987 to 1997. He did his Internal Medicine residency at the Medical College of Wisconsin in 1989–1992. He did his Hematology/Oncology and Bone Marrow Transplant Fellowship at the University of Washington, Fred Hutchinson Cancer Research Center (FHCRC) in Seattle, from 1992 to 1997 and trained under Professors Don Thomas (Nobel laureate), Dean Buckner, Frederick Appelbaum, and Rainer Storb in the clinical division. He is a diplomate of the American Board of Internal Medicine and board certified in Internal Medicine and Medical Oncology. He was given the title of 'Fellow of American College of Physicians' (FACP) by ACP in July 1996. He has conducted many clinical studies at the FHCRC as principal investigator or co-investigator, mainly related to peripheral blood stem cell mobilization and high-dose chemotherapy (HDC) in patients with solid tumors and hematologic malignancies. During his career, he has written many papers in respected medical journals and books with regard to stem cell mobilization kinetics, factors influencing stem collection and engraftment, as well as HDC in patients with lymphoma, multiple myeloma, and breast and ovarian cancer. He was chairman of the EBMT Solid Tumors Working Party (STWP) from 2001 to 2007. He has conducted, chaired, and published studies related to HDC in patients with solid tumors at the EBMT STWP. He was the president of the 29th and 41st EBMT Annual Meetings which were held in Istanbul in 2003 and 2015. He has been a member of the editorial board of several journals. Dr. Demirer currently serves as professor of Medicine and Hematology/Oncology at Ankara University Medical School, Turkey, and as a full member of the Turkish Academy of Sciences as well as the European Academy of Sciences and Arts (EASA).

Contents

Preface

This book covers a wide range of issues related to new developments and innovations in cell-based therapies discussed in basic and clinical chapters from authors around the world involved in stem cell studies and research. The book documents the increased amount of stem cell-related research, basic and clinical applications as well as views for the future. It thereby complements and extends the basic coverage of stem cells, such as mesenchymal stem cells, effect of stem cells on aging, cover hematopoietic stem cells, storage and cryopreservation, issues related to clinical applications such as haploidentical transplants and use of stem cells for the treatment of Huntington's disease. Clearly, developments in the treatment of various malignant and nonmalignant diseases and biomedical engineering depend heavily on stem cells, and this book is well positioned to provide comprehensive coverage for some of these developments.

With the increased number of publications related to stem cells in Cell Transplantation, we consider it important and take this opportunity to share these new developments and innovations in stem cell research and the cell transplantation field with our worldwide readers. Therefore, I can say that this book will be the main source for clinical and preclinical publications for scientists working toward cell transplantation therapies with the primary goal of replacing diseased cells with donor cells of various organs and transplanting those cells close to the injured or diseased target.

Stem cells have a unique ability: they are able to self-renew limitlessly, which allows them to replenish themselves as well as other cells. Another ability of stem cells is that they are able to differentiate into any cell type. A stem cell does not differentiate directly into a specialized cell; there are often multiple intermediate stages. A stem cell first differentiates into a progenitor cell; a progenitor cell is similar to a stem cell, although they are limited in the number of times they can replicate and they are also restricted in which cells they can further differentiate into. Serving as a sort of repair system for the body, they can theoretically divide without limit to replenish other cells as long as the person or animal is still alive. When a stem cell divides, each new cell has the potential to either remain a stem cell or become another type of cell with a more specialized function, such as a muscle cell, a red blood cell or a brain cell.

It will be exciting and interesting for our readers to follow the recent developments in the field of basic and clinical aspects of stem cells and cell transplantation. Although we are close to finding pathways for stem cell therapies in many medical conditions, scientists need to be careful about how they use stem cells ethically and should not rush into clinical trials without carefully investigating the side effects. The focus must be on Good Manufacturing

Practices (GMP) and careful monitoring of the long-term effects of transplanted stem cells in the host.

During the last decade, the number of published articles or books investigating the role of stem cells in cell transplantation or regenerative medicine increased remarkably across all sections of stem cell-related journals. The largest number of stem cell-related articles was published mainly in the fields of clinical transplantation and neuroscience, followed by the bone, muscle, cartilage and hepatocytes. Interestingly, in recent years, the number of stem cell-related articles describing the potential use of stem cell therapy and islet transplantation in diabetes is also slowly increasing, even though this field of endeavor could have one of the greatest clinical and societal impacts.

In conclusion, *Cell Transplantation* is bridging cell transplantation research in a multitude of disease models as methods and technology continue to be refined. The use of stem cells in many therapeutic areas will bring hope to many patients awaiting replacement of malfunctioning organs or repair of damaged tissue. We hope that this book will be an important tool and reference guide for all scientists worldwide who work in the field of stem cells and cell transplantation and that it will shed light upon many important debatable issues in this field.

I would like to thank all the authors who contributed to this book with excellent and up-to-date chapters relaying the recent developments to our readers in the field of stem cell transplantation. I would like to give a special thanks to Iva Simcic, Publishing Process Manager, and all the staff at Intech for their valuable contribution in order to make this book possible.

Taner DEMİRER, MD, FACP
Professor of Medicine, Hematology/Oncology
Dept. of Hematology
Ankara University Medical School
Ankara, TURKEY

Chapter 1

Hematopoietic Stem Cell Source and Storage

Sinem Civriz Bozdag and Osman İlhan

Additional information is available at the end of the chapter

http://dx.doi.org/10.5772/60994

Abstract

Hematopoietic stem cell transplantation(HSCT), has been accepted as a feasible treatment option that prolongs survival in hematological malignancies. Stem cell choice during hematopoietic stem cell transplantation can differ according to the experience of physicians, mostly treated hematological diseases in the centers or ongoing clinical trials. In this chapter we will discuss the advantages and disadvantages of three stem cell sources peripheral blood, bone marrow and umbilical cord blood.

Keywords: hematopoietic stem cell

1. Introduction

Hematopoietic stem cell transplantation (HSCT) has been accepted as a feasible treatment option that prolongs survival in hematological malignancies. Bone marrow (BM) has been widely used as stem cell source in the early stem cell transplantation series. Tendency towards peripheral blood (PB) as stem cell source has been started in the beginning of 2000s. Initially, advantages of peripheral blood stem cell collection have been demonstrated in autologous stem cell transplantation (autoSCT). The introduction of peripheral blood into allogeneic stem cell transplantation (alloSCT) has been followed by allogeneic transplantation from unrelated donors. Due to the less stringent HLA matching requirement, cord blood stands out as an option for patients who do not have HLA-matched donors.

Stem cell choice can change according to the experience of physicians, mostly treated hematological diseases in the centers or ongoing clinical trials. In this chapter, we will discuss the

INTECH
open science | open minds

© 2015 The Author(s). Licensee InTech. This chapter is distributed under the terms of the Creative Commons Attribution License (http://creativecommons.org/licenses/by/3.0), which permits unrestricted use, distribution, and reproduction in any medium, provided the original work is properly cited.

advantages and disadvantages of three stem cell sources: peripheral blood, bone marrow, and umbilical cord blood.

2. Peripheral blood versus bone marrow

Although the early transplantation series mostly used bone marrow as stem cell source, the administration of granulocyte colony-stimulating factor allowed physicians to collect stem cells from peripheral blood by apheresis procedure. Studies in autologous stem cell transplantation (autoSCT) have been reported first. Faucher et al. [1,2] compared peripheral blood and bone marrow in a series with small patient numbers and reported an improvement in hematological recovery and a decrease in hospitalization duration. Survival of the patients was found to be similar between these two stem cell sources [3]. Also, the cost-effectiveness of the procedure increased peripheral blood stem cell transplantation rates compared with bone marrow transplantation. In a study, the total cost in peripheral blood stem cell transplantation was found to be 27.5% less than bone marrow transplantation. Faster hematological recovery, fewer hospitalization days, and less antibiotic treatment created this difference in total cost [4].

Afterwards, the comparison of peripheral blood with bone marrow in allogeneic stem cell transplantation (alloSCT) setting was started. Bensinger et al. [5] reported faster hematological recovery, less transfusion numbers, less severe acute graft versus host disease (a gvhd in favor of PB), but similar chronic graft versus host disease (cGVHD) rates. Our centers' experience also revealed faster engraftment, fewer transfusions with PB [6]. Miflin et al. [7] showed that increasing the infused CD34+ cell number over 4× $10e^6$/kg can significantly accelerate the engraftment kinetics.

In a multicenter randomized trial performed by the European Group for Blood and Marrow Transplantation (EBMT), transplant-related mortality and leukemia-free survival rates showed no significant difference between PB and BM [8]. Mielcarek et al. [9] reported a better 10 years disease-free survival in favor of PB but similar overall survival between two stem cell sources. In a Cochrane database review, trials including related stem cell donors were analyzed. Both neutrophil and platelet recoveries were faster in PB. Disease-free survival, nonrelapse mortality was not different between PB and BM. The advantage of relapse in PB was recorded [10].

Peripheral blood has more CD34+ cells but has also more T cells in comparison with BM. This reflects to the outcomes as more acute and chronic GVHD rates. In a study, acute grades I–IV GVHD incidences for infused CD34+ cell doses less than 2× $10e^6$/kg, between 2× $10e^6$/kg and 4× $10e^6$/kg, and more than 4× $10e^6$/kg were 21%, 35%, and 43%, respectively. In the same study, increase in CD3+ cell dose was also identified as an independent factor for acute GVHD [11]. The correlation between infused cell dose and chronic GVHD has also been shown. Zaucha et al. reported CD34+ cell doses over 8× $10e^6$/kg were found to be associated with increased clinical extensive chronic GVHD. However, this association could not be shown with stem cell transplantation from bone marrow [12]. cGVHD after PB transplantation has to be treated for a longer period with higher immunosuppressive regimens. The presence of cGVHD has been found to be related with fewer relapses but more treatment-related mortality [13–16].

The comparison of stem cell sources has been performed according to the diagnosis of hematological malignancies. Pidala et al. [17] performed a Markov model in which PB was found to be the optimum stem cell source for hematological malignancies, which had an advantage of 7 months in comparison with BM. BM was chosen to be superior in patients with 1-year relapse rates lower than 0.05. Patients with high-risk hematological malignancies like acute leukemia in second or later remission, chronic myeloid leukemia (CML) in blastic transformation, refractory anemia with excess blast in transformation, and heavily pretreated lymphoma patients were found to benefit from PB transplants in comparison with BM transplants [16]. A meta-analysis of nine randomized clinical trials showed better disease-free and overall survival with PB in late stage disease [18]. In a randomized study of patients with myeloid malignancies, acute myeloid leukemia AML,CML and myelodysplastic syndrome (MDS), hematological recovery was faster in PB group. Improvement in overall survival with PB was found to be related with reduction in nonrelapse deaths, with no difference in early, late relapses, or deaths in relapses [19]. In another study, which has included chronic myeloid leukemia patients, the incidence of acute and extensive chronic GVHD was similar between BM and PB patients, and there was no significant difference between survival and nonrelapse mortality rates. However, in the subgroup analysis of chronic phase patients, PB patients experienced more chronic GVHD, and BM patients had higher relapse rates [20]. The advantage of BM remains in benign hematological diseases, where chronic GVHD rates affect transplant outcomes negatively. In a study that included 537 adolescent aplastic anemia patients who had alloSCT, the survival advantage for BM recipients was reported to be significant [21]. Bacigalupo et al. [22] showed the advantage of BM also for the patients older than 50 years. The major causes of death were GVHD, infections, and graft rejection in this study. GCSF-stimulated and GCSF-unstimulated bone marrow and peripheral blood have been compared in another study that included aplastic anemia patients. Engraftment rates were not different in three treatment arms. Grades II–IV acute GVHD and chronic GVHD rates were higher with PB transplants in comparison with BM. GCSF-manipulated BM was not superior to BM in terms of mortality rates [23].

Late effects of transplantation were also analyzed in studies. Performance status, return to work, incidence of bronchiolitis obliterans, and hematopoietic functions were found to be similar between PB and BM. Also, there was no significant difference in secondary malignancies between the groups [24].

As HLA-matched siblings can be found in 25% of the patients, either PB or BM of the unrelated donors has been accepted as sources for stem cell transplantation. Eapen et al [25] reported significantly higher acute and chronic GVHD risk but similar transplant-related mortality and survival rates with PB than BM in unrelated allogeneic stem cell transplants. In a phase 3, multicenter randomized trial from 46 transplant centers in the United States and Canada, only chronic GVHD rates were significantly higher in PB transplantations. In long-term outcome analysis, chronic GVHD was graded as extensive in 85% of PB recipients compared with 76% of BM recipients. Neutrophile and platelet engraftment rates were found to be significantly higher in PB transplantations, but no difference was observed in survival outcomes [26]. In a Cochrane database review, trials including unrelated stem cell donors were also revised. Both neutrophile and platelet recovery was faster in PB transplantations. Disease-free survival, nonrelapse mortality of PB, and BM recipients were not different. The relapse advantage of PB was not proved. Acute grades II–IV incidence did not reveal a statistical significance, whereas extensive chronic GVHD rates were in favor of BM [10].

The outcomes of unrelated allogeneic stem cell transplantation were also separately assessed according to the diagnosis of hematological diseases. During an overall survival of 7 years, nonrelapse mortality and relapse rates were similar in acute myeloid leukemia and myelodysplastic syndrome patients who have received PB or BM. The only significant difference was reported in chronic myeloid leukemia patients due to higher nonrelapse mortality rates in PB transplants [27]. In a Korean study, the difference in GVHD incidence could be overcomed by risk adapted GVHD prophylaxis in AML patients [28]. In aplastic anemia patients, the risk of death was higher with unrelated donors, but peripheral blood as a stem cell source was a negative predictor for outcomes [29].

The impact of conditioning regimens in unrelated allogeneic stem cell transplantation has also been assessed in clinical trials. EBMT data showed that when reduced intensity conditioning (RIC) regimens were used in unrelated transplant settings, acute and chronic GVHD rates were higher, and relapse rates were lower in PB transplants of AML patients [30].In contradictory with these results, GVHD, relapse, and survival rates of hematological malignancies who had unrelated stem cell transplantation were reported to be similar in a recent trial [31].

The issues that have to be paid importance during hematopoietic stem cell transplantation should be the benefits of either the donor or the patient.

Both the peripheral blood and the bone marrow have advantages and disadvantages for the donors. Peripheral blood donation is a collection of HSC from peripheral blood after a 5-day course of granulocyte colony-stimulating factor administration on 1–2 days via 4–5 hours of apheresis procedure. The monitorization of circulating CD34+ cells on the first day of apheresis is predictive for stem cell yield [32].

Peripheral blood stem cell collection seems to be an easier collection method, but the central venous line can be a necessity for some of the donors if the standard peripheral venous line is not adequate. Again in NMDP experience, femoral and jugular lines were in same frequency, which was twice as the subclavian line [33]. The collection of peripheral blood stem cells via apheresis procedure has been proven to be an effective and safe method for donors [34]. Bone marrow harvesting is the collection of HSC from posterior iliac crest under anesthesia. Hospital admission can be necessary for postoperative follow-up. In a prospective study, the experience of donors for bone marrow and peripheral blood collection has been compared. Bone marrow donors were found to be less confused and more prepared for donation. They also found the process psychologically beneficial in a short term. However, the long-term health-related quality was similar between both donors [35]. In a survey of 51,024 AHSCT performed by 338 teams, five donor fatalities were observed. Severe adverse events were reported in 37 donors. Hematological malignancy rates were not different from the age and sex-adjusted general population [36]. Bone pain was the most frequent side effect of donors; in bone marrow donors, pain at the collection site and in peripheral blood donors pain at various sites of body during GCSF administration have been reported. Tiredness, light-headedness, nausea, sleeping problems, and chills were the other frequent side effects. At 12 months postdonation, the most common side effects were tiredness and muscle aches [35]. A median time to recovery in PB and BM donors were 1 and 2 weeks, respectively [37]. In a prospective trial from NMDP, females and heavier donors had more III–IV CALGB adverse events [38].

3. Umbilical cord blood

Umbilical cord remains to be a stem cell source option for patients who do not have a matched sibling or unrelated donor. The first umbilical cord blood transplantation has been performed at the end of the 1980s in a child diagnosed with Fanconi anemia [39]. Today, the storage of more than 600,000 cryopreserved cord bloods serves as an alternative option for both pediatric and adult patients. In the U.S. registry, almost all of the patients younger than 20 years and 80% of the patients older than 20 years had cord blood mismatched units in one or two HLA locus [40].

Rapid availability, no requirement for full HLA match, and being an option for ethnic minorities became the advantages of umbilical cord. Also, no harm for donor can be another important feature in stem cell source choice. Lack of sufficient cell doses for adult patients, delayed engraftment, poor immune reconstitution, and high infection rates are the major disadvantages of cord blood transplantation. Cord blood includes more naive T cells and Tregs. The first naive T cells proliferate but show a limited repertoire. Then a thymic-dependent population expands, which can be affected by conditioning regimen, GVHD, or aging [41].

The standard practice for the HLA typing of cord blood units is to analyze A-B antigens and DRB1 alleles with high resolution. The acceptable HLA match is 4 to 6/6 match for performing transplantation, but each mismatch results in increased TRM, increased severity of acute GVHD, and decreased survival [42]. Eapen et al. [43] reported the impact of level matching on TRM and neutrophile engraftment. Neutrophile recovery was found to be delayed in transplantations with more than two mismatches, and nonrelapse mortality was reported to be higher in 1–5 mismatch compared with HLA full-match transplantations. HLA C matching has also been shown to be beneficial in studies [42,44]. Barker et al. [45] analyzed 1691 MDS and AML patients who received single cord blood transplantation and found that regardless of cell dose, HLA A, B, and DR-matched transplantations result in the best outcomes. Total nucleated cell (TNC) doses greater than 2.5×10^{e7}/kg in one mismatch recipient and $5\times 10e^{7}$/kg in two mismatch recipients have been shown to be sufficient for better survival. In another study, high-resolution DRB1 match was related with less acute GVHD and better event-free survival rates. Also, in the same study, infusing higher CD34+ cells, CD34+HLA DR+ CD38+ cells, and CD3+ cells resulted in faster engraftment [46]. The impact of HLA mismatch direction between donor and recipient has also been assessed by EUROCORD/EBMT analysis. Neither one to two mismatch in graft versus host direction nor host versus graft direction was found to be related with increased nonrelapse mortality and survival [47].

Graft failure risk remains to be a problem in 10%–20% of patients who had cord blood transplantation [48]. Neutrophile and platelet engraftment is delayed in cord blood transplantations. Engraftment is shown to correlate with cell dose infused [49]. In an analysis of 1268 patients with acute leukemia, 3 years overall survival was 47% and TRM was 16%. Delay in engraftment was associated with increased mortality and shorter survival rates [50].

Survival outcomes have been improved over years with better patient, conditioning regimen selection, and progress in HLA typing [44]. Rubinstein et al. [51] showed that 46% of the patients experienced transplant-related events by posttransplant day 100. Transplant-related events and event-free survival were related with diagnosis, number of leukocytes in the transplant, age, extent of HLA disparity, and transplant center. Cohen et al. [52] reported the outcome of 500 patients who had single-unit cord blood transplantation from 1995 to 2005 and found 1 year survival as 37%. Factors affecting early mortality following the myeloablative single-unit cord blood transplantation were cell dose, advanced disease, older age, cytomegalovirus status, female gender, and limited cord blood center experience. A Japanese study revealed that disease status and cytogenetics had an impact on event-free survival rates in AML patients [53]. For acute lymphoblastic leukemia patients who had cord blood transplantation, the factors associated with better leukemia-free survival were age, advanced disease, and conditioning regimen [54]. Brunstein et al. [55] reported 3 years event-free and overall survival as 38% and 45%, respectively, by performing transplantation with nonmyeloablative conditioning. In a comparative study, on the effect of conditioning regimen intensity, transplant-related mortality was similar with both regimens; lower risk of relapse and longer leukemia-free survival could be achieved after myeloablative regimens [56]. Fludarabine in combination with total body irradiation (TBI) concluded with high treatment-related mortality [57]. In another trial, fludarabine, TBI in combination with busulphan versus cyclophosphamide has been investigated; cyclophosphamide resulted in better transplantation outcome [58]. Also, the Japanese group used a myeloablative conditioning regimen, which included TBI, cytarabine, and cyclophosphamide, and reported 51% overall survival for high-risk hematological malignancies [59].

Graft versus host disease (GVHD) is one of the major early and late morbidity and mortality causes. It has been proven that donor source has an impact on GVHD rates. Although the HLA disparities were higher in cord blood transplantations, GVHD was found to be lower than bone marrow [60–62]. Eurocord and EBMT revealed that HLA mismatch increased the acute GVHD risk in cord blood transplantation [63]. Patient age [51], CMV status [63], nonmyeloablative conditioning, and absence of ATG were also factors thought to be related with acute GVHD. Acute GVHD was reported to be higher in double-unit cord transplants than the single-unit transplants [60]. In a study with 1072 patients, chronic GVHD in posttransplant 2 years was 28%. In multivariate analysis, risk factors were identified as myeloablative conditioning regimen, mycophenolate mofetil in GVHD prophylaxis, increased HLA mismatch, higher body weight, and previous acute GVHD [64]. Newell et al. [65] found higher chronic GVHD rates, which were analyzed according to NIH 2005 criteria and had a predominance of acute GVHD features.

Comparative trials of umbilical cord blood with different stem cell sources have been reported. Neutrophile and platelet recovery have been delayed in cord blood transplantations, and chronic GVHD rates have been observed less. Laughlin et al. [66] included patients who had received either an HLA-matched marrow transplant- a marrow transplant with a single HLA mismatch from an unrelated donor or who had received a cord blood transplant with one or two HLA mismatches in their study. Overall mortality was found to be lower in matched related BM recipients. The rate of leukemia recurrence was found to be similar, but 3 years survival for cord blood recipients was 26%, which was lower than matched BM recipients' survival. In a recent trial, donor types have been investigated in AML patients; survival rates were similar between matched related, matched unrelated, mismatched unrelated, and cord blood transplant. Age and type of conditioning regimens were the major determinants of survival [67].

As a consequence of inadequate engraftment with single-unit cord blood transplantation in adult hematological malignancy patients, double-unit cord transplantation was introduced in the beginning of 2000s. Sustained engraftment could be achieved more than 90% of patients and one of the double-units dominate [68,69]. Barker et al. [70] observed the impact of CD3+ cell dose on engraftment, but not the CD34+ cell dose or HLA match in double cord unit transplantations.

Avery et al. [68] found an association between higher numbers of CD34+ and TNC cell dose of the dominant unit and sustained engraftment. In the same trial, unit–unit HLA match and unit–recipient HLA match were not associated with sustained engraftment. Mixed chimerism can be displayed in follow-up, especially if the HLA match of the both units is close [71]. The HOVON group showed that the unit predominance was observed by posttransplant day 11, and the role of CD4+ lymphocyte-mediated alloreactivity was suggested [72]. In another study, the cord blood bankreported pre cryopreservation and postthaw viable CD34+ cell doses were the important parameters for the engraftment [73].

Macmillan et al. [60] reported double cord transplantation as one of the risk factors for acute GVHD. Ponce et al. [74] demonstrated the acute grades II–IV rates as 53% and predominantly the gut as affected organ. Chronic GVHD rates have been reported around 30% in double cord blood transplantation trials [75,76].

Eurocord and EBMT compared single-unit transplantation with different myeloablative conditioning regimens and double-unit transplantation. Conditioning regimens were either the widely accepted Minnesota protocol, which consisted of TBI/cyclophosphamide/fludarabine or thiotepa/busulphan/fludarabine (TBF). In this study, 2 years LFS was similar between double-unit cord and single-unit cord if the new TBF protocol was chosen [75,77]. In a prospective multicenter trial, which included 56 acute leukemia and myelodysplasia patients transplanted with double-unit cord, 3 years disease-free survival was 50% and TRM was 39% [76].

Comparative studies of different stem cell sources in the era of double-unit cord have been continued to be reported. In comparison with filgrastim mobilized peripheral blood, delay in immune reconstitution of T cells in cord blood transplantation resulted in increased infection risks. The double-unit cord blood has been compared with unrelated donor grafts; although 3 years survival analysis were similar, double cord blood was associated with less chronic GVHD but more nonrelapse mortality [78]. Brunstein et al. [79] have found comparable 5-year leukemia-free survival after HLA-matched related, unrelated, and double-unit cord blood transplantation.

4. Hematopoietic progenitor cell and umbilical cord blood cryopreservation

Collected hematopoietic stem cells have been cryopreserved by using a cryoprotectant, dimethyl sulfoxide (DMSO), and frozen in liquid nitrogen vapor until reinfusion. DMSO penetrates to cells and binds to water molecules in order to prevent dehydration of the cells [80]. During the cryopreservation process, DMSO dilution, freezing period, storage in vapor phase, or liquid nitrogen are the important factors for optimum results. DMSO and cell suspension should be cooled down to 0°C–4°C, and after the addition of DMSO, the product should be placed in a controlled freezer subsequently. The optimum concentration of DMSO has also been analyzed in different studies because it can be toxic for stem cell viability and also may cause side effects during infusion. Although majority of transplant centers still prefer to use 10% DMSO, lower percentage of DMSO or washing the product before infusion to decrease toxicity has also been used [81]. Reducing DMSO concentration into 7.5% has been revealed as feasible [82]. Up to 10 years cryopreservation with 5% DMSO has also been found not to have a negative impact on cell viability [83]. DMSO concentration has been calculated as ml/kg body weight or ml/min, and in EBMT results from 65 centers, it has been revealed that calculation as milliliters per minute should be the preferred way to reduce side effects [84]. DMSO toxicity can result in nausea, vomiting, fever, or more severe reactions like hepatic dysfunction, cardiac arrhythmia, and neurotoxicity [85,86].

In the freezing period, cells have been mostly frozen using controlled rate freezers. Too fast cooling can result in intracellular crystallization, and too slow cooling can induce extracellular ice formation. Sputtek et al. [87] reported that the cooling rate range can vary from 1 to 5 K/min. Also, the recovery of white blood cell (WBC) recovery was found to be superior in slow rate freezing to fast rate freezing [88]. After the introduction of stem cell storage in liquid nitrogen, risks for microbial contamination of the products concluded with the usage of vapor phase for storage. However, the comparison of the two phase was found to be similar for either WBC recovery or WBC viability in comparative studies [88,89].

The major disadvantage of the umbilical cord blood processing is the potential risk for loss of progenitor cells in the collected product. The techniques that have been used for red cell separation like simple centrifugation lysis with ammonium chloride, filtration through density gradients, or collection from bags to vessels were found to have detrimental effects during cryopreservation [90–93]. Thus, Rubinstein et al. [94] initially proved that forming 20 ml cord blood units with uniform volume can be achieved by using rouleaux formation induced by hydroxyethyl starch and centrifugation. Semiautomated top–bottom systems and lately

http://dx.doi.org/10.5772/60994

automatic devices like AXP-SEPAX have been developed [95]. Automatic systems could achieve similar cell recovery with less technical influence [96].

After the addition of 10% dimethyl sulfoxide, cord blood samples can be cooled from 4°C to –80°C mostly by controlled rate freezers [97]. It has been recently shown that cord bloods can be transferred into liquid nitrogen vapor phase directly or after storage at –80°C for 18 hours [98]. It has been demonstrated by Broxmeyer et al. that the long-term storage does not have a negative influence on in vitro function of umbilical cord blood progenitor cells. Also, the duration of cryopreservation was found to have no impact on clinical outcome like neutrophile or platelet recovery after cord blood transplantation [99].

5. Conclusion

Hematopoietic stem cell source choice is an important issue to be concerned during stem cell transplantation. Diagnosis and pretransplantation status of hematological diseases and type of conditioning regimens are the major factors in making decision for one type of stem cell source.

Processing before the storage of collected stem cells can show differences according to the source of stem cells. In particular, cord blood processing needs more attention due to the risk of hematopoietic stem cell loss.

Author details

Sinem Civriz Bozdag* and Osman İlhan*

*Address all correspondence to: osmanilhan@medicine.ankara.edu.tr

Stem Cell Transplantation Unit, Department of Hematology, School of Medicine, Ankara University, Ankara, Turkey

References

[1] Faucher C, le Corroller AG, Blaise D, Novakovitch G, Manonni P, Moatti JP, Maraninchi D. Comparison of G-CSF-primed peripheral blood progenitor cells and bone marrow auto transplantation: clinical assessment and cost-effectiveness. Bone Marrow Transplant. 1994 Dec;14(6):895–901.

[2] Faucher C, Le Corroller AG, Chabannon C, Viens P, Stoppa AM, Bouabdallah R, Camerlo J, Vey N, Gravis G, Gastaut JA, Novakovitch G, Mannoni P, Bardou VJ, Moatti JP, Maraninchi D, Blaise D. Autologous transplantation of blood stem cells mobilized

with filgrastim alone in 93 patients with malignancies: the number of CD34+ cells re-infused is the only factor predicting both granulocyte and platelet recovery. J Hematother. 1996 Dec;5(6):663–70.

[3] Bensinger WI, Weaver CH, Appelbaum FR, Rowley S, Demirer T, Sanders J, Storb R, Buckner CD. Transplantation of allogeneic peripheral blood stem cells mobilized by recombinant human granulocyte colony-stimulating factor. Blood. 1995 Mar 15;85(6): 1655–8.

[4] Duncan N, Hewetson M, Powles R, Raje N, Mehta J. An economic evaluation of peripheral blood stem cell transplantation as an alternative to autologous bone marrow transplantation in multiple myeloma. Bone Marrow Transplant. 1996 Dec;18(6):1175–8.

[5] Bensinger WI, Clift R, Martin P, Appelbaum FR, Demirer T, Gooley T, Lilleby K, Rowley S, Sanders J, Storb R, Buckner CD. Allogeneic peripheral blood stem cell transplantation in patients with advanced hematologic malignancies: a retrospective comparison with marrow transplantation. Blood. 1996 Oct 1;88(7):2794–800.

[6] Ustün C, Arslan O, Beksaç M, Koç H, Gürman G, Ozçelik T, Yilmaz B, Ilhan O, Akan H, Ozcan M, Demirer T, Uysal A, Konuk N, Arat M, Dilek I, Celebi H, Coskun HS. A retrospective comparison of allogeneic peripheral blood stem cell and bone marrow transplantation results from a single center: a focus on the incidence of graft-vs.-host disease and relapse. Biol Blood Marrow Transplant. 1999;5(1):28–35.

[7] Miflin G, Russell NH, Hutchinson RM, Morgan G, Potter M, Pagliuca A, Marsh J, Bell A, Milligan D, Lumley M, Cook G, Franklin Allogeneic peripheral blood stem cell transplantation for haematological malignancies—an analysis of kinetics of engraftment and GVHD risk. Bone Marrow Transplant. 1997 Jan;19(1):9–13.

[8] Schmitz N, Bacigalupo A, Hasenclever D, Nagler A, Gluckman E, Clark P, Bourquelot P, Greinix H, Frickhofen N, Ringdén O, Zander A, Apperley JF, Gorin C, Borkett K, Schwab G, Goebel M, Russell NH, Gratwohl A. Allogeneic bone marrow transplantation vs filgrastim-mobilised peripheral blood progenitor cell transplantation in patients with early leukaemia: first results of a randomised multicentre trial of the European Group for Bloodand Marrow Transplantation. Bone Marrow Transplant. 1998 May;21(10):995–1003.

[9] Mielcarek M, Storer B, Martin PJ, Forman SJ, Negrin RS, Flowers ME, Inamoto Y, Chauncey TR, Storb R, Appelbaum FR, Bensinger WI. Long-term outcomes after transplantation of HLA-identical related G-CSF-mobilized peripheral blood mononuclear cells versus bone marrow. Blood. 2012 Mar 15;119(11):2675–8.

[10] Holtick U, Albrecht M, Chemnitz JM, Theurich S, Skoetz N, Scheid C, von Bergwelt-Baildon M. Bone marrow versus peripheral blood allogeneic haematopoietic stem cell transplantation for haematological malignancies in adults. Cochrane Database Syst Rev. 2014 Apr 20;4:CD010189.

[11] Urbano-Ispizua A, Rozman C, Pimentel P, Solano C, de la Rubia J, Brunet S, Pérez-Oteyza J, Ferrá C, Zuazu J, Caballero D, Bargay J, Carvalhais A, Díez JL, Espigado I, Alegre A, Rovira M, Campilho F, Odriozola J, Sanz MA, Sierra J, García-Conde J, Montserrat E; Spanish Group for AllogeneicPeripheral Blood Transplantation (Grupo Español de Trasplante Hemopoyético) and Instituto Português de Oncologia-Porto. Risk factors for acute graft-versus-host disease in patients undergoing transplantation with CD34+ selected blood cells from HLA-identical siblings. Blood. 2002 Jul 15;100(2):724–7.

[12] Zaucha JM, Gooley T, Bensinger WI, Heimfeld S, Chauncey TR, Zaucha R, Martin PJ, Flowers ME, Storek J, Georges G, Storb R, Torok-Storb B. CD34 cell dose in granulocyte colony-stimulating factor-mobilized peripheral blood mononuclear cell grafts affects engraftment kinetics and development of extensive chronic graft-versus-host disease after human leukocyte antigen-identical sibling transplantation. Blood. 2001 Dec 1;98(12):3221–7.

[13] Urbano-Ispìzua A. Chronic graft versus host disease after allogeneic peripheral blood transplantation. Haematologica. 2003 Mar;88(3):245–6.

[14] Lee SJ. Have we made progress in the management of chronic graft-vs-host disease? Best Pract Res Clin Haematol. 2010 Dec;23(4):529–35.

[15] Boyiadzis M, Arora M, Klein J, Hassebroek A, Hemmer M, Urbano-Ispizua A, Antin JH, Bolwell B, Cahn JY, Cairo MS, Cutler C, Flowers M, Gale R, Herzig R, Isola L, Jacobsohn D, Jagasia M, Klumpp T, Lee SJ, Petersdorf E, Santarone S, Spellman S, Schouten HC, Verdonck L, Wingard J, Weisdorf D, Horowitz M, Pavletic S. Impact of chronic graft-versus-host disease on late relapse and survival on 7489 patients after myeloablative allogeneic hematopoietic cell transplantation for leukemia. Clin Cancer Res. 2014 Oct 27.

[16] Cheuk DK. Optimal stem cell source for allogeneic stem cell transplantation for hematological malignancies. World J Transplant. 2013 Dec 24;3(4):99–112.

[17] Pidala J, Anasetti C, Kharfan-Dabaja MA, Cutler C, Sheldon A, Djulbegovic B. Decision analysis of peripheral blood versus bone marrow hematopoietic stem cells for allogeneic hematopoietic cell transplantation. Biol Blood Marrow Transplant. 2009 Nov;15(11):1415–21.

[18] Stem Cell Trialists' Collaborative Group. Allogeneic peripheral blood stem-cell compared with bone marrow transplantation in the management of hematologic malignancies: an individual patient data meta-analysis of nine randomized trials. J Clin Oncol. 2005; 23: 5074–87.

[19] Couban S, Simpson DR, Barnett MJ, Bredeson C, Hubesch L, Howson-Jan K, Shore TB, Walker IR, Browett P, Messner HA, Panzarella T, Lipton JH; Canadian Bone Marrow Transplant Group. A randomized multicenter comparison of bone marrow and

peripheral blood in recipients of matched sibling allogeneic transplants for myeloid malignancies. Blood. 2002 Sep 1;100(5):1525–31.

[20] Oehler VG, Radich JP, Storer B, Blume KG, Chauncey T, Clift R, Snyder DS, Forman SJ, Flowers ME, Martin P, Guthrie KA, Negrin RS, Appelbaum FR, Bensinger W. Randomized trial of allogeneic related bone marrow transplantation versus peripheral blood stem cell transplantation for chronic myeloid leukemia. Biol Blood Marrow Transplant. 2005 Feb;11(2):85–92.

[21] Dufour C, Pillon M, Passweg J, Socié G, Bacigalupo A, Franceschetto G, Carraro E, Oneto R, Risitano AM, Peffault de Latour R, Tichelli A, Rovo A, Peters C, Hoechsmann B, Samarasinghe S, Kulasekararaj AG, Schrezenmeier H, Aljurf M, Marsh J. Outcome of aplastic anemia in adolescence: a survey of the Severe Aplastic Anemia Working Party of the European Group for Blood and Marrow Transplantation. Haematologica. 2014 Oct;99(10):1574–81.

[22] Bacigalupo A, Socié G, Schrezenmeier H, Tichelli A, Locasciulli A, Fuehrer M, Risitano AM, Dufour C, Passweg JR, Oneto R, Aljurf M, Flynn C, Mialou V, Hamladji RM, Marsh JC; Aplastic Anemia Working Party of the European Group for Blood and Marrow Transplantation (WPSAA-EBMT). Bone marrow versus peripheral blood as the stem cell source for sibling transplants in acquired aplastic anemia: survival advantage for bone marrow in all age groups. Haematologica. 2012 Aug;97(8):1142–8.

[23] Chu R, Brazauskas R, Kan F, Bashey A, Bredeson C, Camitta B, Chiang KY, Frangoul H, Gale RP, Gee A, George B, Goldman FD, Gross TG, Gupta V, Hale GA, Isola L, Ispizua AU, Lazarus H, Marsh J, Russell J, Sabloff M, Waller EK, Eapen M. Comparison of outcomes after transplantation of G-CSF-stimulated bone marrow grafts versus bone marrow or peripheral blood grafts from HLA-matched sibling donors for patients with severeaplastic anemia. Biol Blood Marrow Transplant. 2011 Jul;17(7): 1018–24.

[24] Friedrichs B, Tichelli A, Bacigalupo A, Russell NH, Ruutu T, Shapira MY, Beksac M, Hasenclever D, Socié G, Schmitz N. Long-term outcome and late effects in patients transplanted with mobilised blood or bone marrow: a randomised trial. Lancet Oncol. 2010 Apr;11(4):331–8.

[25] Eapen M, Logan BR, Confer DL, Haagenson M, Wagner JE, Weisdorf DJ, Wingard JR, Rowley SD, Stroncek D, Gee AP, Horowitz MM, Anasetti C. Peripheral blood grafts from unrelated donors are associated with increased acute and chronic graft-versus-host disease without improved survival. Biol Blood Marrow Transplant. 2007 Dec;13(12):1461–8.

[26] Anasetti C, Logan BR, Lee SJ, Waller EK, Weisdorf DJ, Wingard JR, Cutler CS, Westervelt P, Woolfrey A, Couban S, Ehninger G, Johnston L, Maziarz RT, Pulsipher MA, Porter DL, Mineishi S, McCarty JM, Khan SP, Anderlini P, Bensinger WI, Leitman SF, Rowley SD, Bredeson C, Carter SL, Horowitz MM, Confer DL; Blood and Marrow

Transplant Clinical Trials Network. Peripheral-blood stem cells versus bone marrow from unrelated donors. N Engl J Med. 2012 Oct 18;367(16):1487–96.

[27] Eapen M, Logan BR, Appelbaum FR, Antin JH, Anasetti C, Couriel DR, Chen J, Maziarz RT, McCarthy PL, Nakamura R, Ratanatharathorn V, Vij R, Champlin RE. Long-term survival after transplantation of unrelated donor peripheral blood or bone marrow hematopoietic cells for hematologic malignancy. Biol Blood Marrow Transplant. 2015 Jan;21(1):55–9.

[28] Shin SH, Kim JH, Jeon YW, Yoon JH, Yahng SA, Lee SE, Cho BS, Eom KS, Kim YJ, Lee S, Min CK, Cho SG, Kim DW, Lee JW, Min WS, Park CW, Kim HJ. Similar outcomes of peripheral blood stem cells vs. bone marrow for human leukocyte antigen-matched unrelated donor transplantation in adult patients with acute myeloid leukemia using risk-adapted graft-versus-host disease prophylaxis. Eur J Haematol. 2014 Jul;93(1):19–28.

[29] Bacigalupo A, Socié G, Hamladji RM, Aljurf M, Maschan A, Kyrcz-Krzemien S, Cybicka A, Sengelov H, Unal A, Beelen D, Locasciulli A, Dufour C, Passweg JR, Oneto R, Signori A, Marsh JC. Current outcome of HLA identical sibling vs. unrelated donor transplants in severe aplastic anemia: an EBMT analysis. Haematologica. 2015 Jan 23.

[30] Nagler A, Labopin M, Shimoni A, Mufti GJ, Cornelissen JJ, Blaise D, Janssen JJ, Milpied N, Vindelov L, Petersen E, Gribben J, Bacigalupo A, Malm C, Niederwieser D, Socié GJ, Arnold R, Brown P, Goker H, Rocha V, Mohty M; ALWP-EBMT. Mobilized peripheral blood stem cells compared with bone marrow from HLA-identical siblings for reduced-intensity conditioning transplantation in acute myeloid leukemia in complete remission: a retrospective analysis from the Acute Leukemia Working Party of EBMT. Eur J Haematol. 2012 Sep;89(3):206–13.

[31] Eapen M, Logan BR, Horowitz MM, Zhong X, Perales MA, Lee SJ, Rocha V, Soiffer RJ, Champlin RE. Bone marrow or peripheral blood for reduced-intensity conditioning unrelated donor transplantation. J Clin Oncol. 2015 Feb 1;33(4):364–9.

[32] Demirer T, Ilhan O, Ayli M, Arat M, Dagli M, Ozcan M, Haznedar R, Genc Y, Fen T, Ayyildiz E, Dincer S, Arslan O, Gurman G, Konuk N, Dalva K, Uysal A, Koc H, Ozet G, Akan H. Monitoring of peripheral blood CD34+ cell counts on the first day of apheresis is highly predictive for efficient CD34+ cell yield. Ther Apher. 2002 Oct; 6(5):384–9.

[33] Bolan CD, Hartzman RJ, Perry EH, Trainor L, Miller J, Miller R, Hanley L, Chitphakdithai P, King RJ. Donation activities and product integrity in unrelated donor allogeneic hematopoietic transplantation: experience of the National Marrow Donor Program. Biol Blood Marrow Transplant. 2008 Sep;14(9 Suppl):23–8.

[34] Arat M, Arslan O, Ayyildiz E, Topcuoglu P, Dalva K, Ilhan O. Peripheral blood stem cell apheresis for allogeneic transplants: Ibni Sina experience. Transfus Apheresis Sci. 2004 Jun;30(3):189–91.

[35] Switzer GE, Bruce JG, Harrington D, Haagenson M, Drexler R, Foley A, Confer D, Bishop M, Anderlini P, Rowley S, Leitman SF, Anasetti C, Wingard JR. Health-related quality of life of bone marrow versus peripheral blood stem cell donors: a prespecified subgroup analysis from a phase III RCT-BMTCTN protocol 0201.Biol Blood Marrow Transplant. 2014 Jan;20(1):118–27.

[36] Halter J1, Kodera Y, Ispizua AU, Greinix HT, Schmitz N, Favre G, Baldomero H, Niederwieser D, Apperley JF, Gratwohl A. Severe events in donors after allogeneic hematopoietic stem cell donation. Haematologica. 2009 Jan;94(1):94–101.

[37] Miller JP, Perry EH, Price TH, Bolan CD Jr, Karanes C, Boyd TM, Chitphakdithai P, King RJ. Recovery and safety profiles of marrow and PBSC donors: experience of the National Marrow Donor Program. Biol Blood Marrow Transplant. 2008 Sep;14(9 Suppl):29–36.

[38] Pulsipher MA, Chitphakdithai P, Logan BR, Navarro WH, Levine JE, Miller JP, Shaw BE, O'Donnell PV, Majhail NS, Confer DL. Lower risk for serious adverse events and no increased risk for cancer after PBSC vs BM donation. Blood. 2014 Jun 5;123(23): 3655–63.

[39] Auerbach AD, Liu Q, Ghosh R, Pollack MS, Douglas GW, Broxmeyer HE. Prenatal identification of potential donors for umbilical cord blood transplantation for Fanconi anemia. Transfusion. 1990;30(8):682–687.

[40] Gragert L, Eapen M, Williams E, Freeman J, Spellman S, Baitty R, Hartzman R, Rizzo JD, Horowitz M, Confer D, Maiers M. HLA match likelihoods for hematopoietic stem-cell grafts in the U.S. registry. N Engl J Med. 2014 Jul 24;371(4):339–48.

[41] Danby R, Rocha V. Improving engraftment and immune reconstitution in umbilical cord blood transplantation. Front Immunol. 2014 Feb 24;5:68.

[42] Eapen M, Klein JP, Sanz GF, Spellman S, Ruggeri A, Anasetti C, Brown M, Champlin RE, Garcia-Lopez J, Hattersely G, Koegler G, Laughlin MJ, Michel G, Nabhan SK, Smith FO, Horowitz MM, Gluckman E, Rocha V; Eurocord-European Group for Blood and Marrow Transplantation; Netcord; Center for International Blood and Marrow Transplant Research. Effect of donor-recipient HLA matching at HLA A, B, C, and DRB1 on outcomes after umbilical-cord blood transplantation for leukaemia and myelodysplastic syndrome: a retrospective analysis. Lancet Oncol. 2011 Dec; 12(13):1214–21.

[43] Eapen M, Klein JP, Ruggeri A, Spellman S, Lee SJ, Anasetti C, Arcese W, Barker JN, Baxter-Lowe LA, Brown M, Fernandez-Vina MA, Freeman J, He W, Iori AP, Horowitz MM, Locatelli F, Marino S, Maiers M, Michel G, Sanz GF, Gluckman E, Rocha V; Center for International Blood and Marrow Transplant Research, Netcord, Eurocord,

and the European Group for Blood and Marrow Transplantation. Impact of allele-level HLA matching on outcomes after myeloablative single unit umbilical cord blood transplantation for hematologic malignancy. Blood. 2014 Jan 2;123(1):133–40.

[44] Ballen KK, Gluckman E, Broxmeyer HE. Umbilical cord blood transplantation: the first 25 years and beyond. Blood. 2013 Jul 25;122(4):491–8.

[45] Barker JN, Scaradavou A, Stevens CE. Combined effect of total nucleated cell dose and HLA match on transplantation outcome in 1061 cord blood recipients with hematologic malignancies. Blood. 2010 Mar 4;115(9):1843–9.

[46] van Heeckeren WJ, Fanning LR, Meyerson HJ, Fu P, Lazarus HM, Cooper BW, Tse WW, Kindwall-Keller TL, Jaroscak J, Finney MR, Fox RM, Solchaga L, Forster M, Creger RJ, Laughlin MJ. Influence of human leucocyte antigen disparity and graft lymphocytes on allogeneic engraftment and survival after umbilical cord blood transplant in adults. Br J Haematol. 2007 Nov;139(3):464–74

[47] Cunha R, Loiseau P, Ruggeri A, Sanz G, Michel G, Paolaiori A, Socié G, Arcese W, Picardi A, Dias de Heredia C, Rio B, Locatelli F, O'Brien TA, Yakoub-Agha I, Angel Diaz M, Milpied N, Bittencourt H, Pedro Souza M, Aljurf M, Charron D, Boudjedir K, Labopin M, Gluckman E, Rocha V. Impact of HLA mismatch direction on outcomes after umbilical cord blood transplantation for hematological malignant disorders: a retrospective Eurocord-EBMT analysis. Bone Marrow Transplant. 2014 Jan; 49(1):24–9.

[48] Kekre N, Antin JH. Hematopoietic stem cell transplantation donor sources in the 21st century: choosing the ideal donor when a perfect match does not exist. Blood. 2014 Jul 17;124(3):334–43.

[49] Migliaccio AR, Adamson JW, Stevens CE, Dobrila NL, Carrier CM, Rubinstein P. Cell dose and speed of engraftment in placental/umbilical cord blood transplantation: graft progenitor cell content is a better predictor than nucleated cell quantity. Blood. 2000 Oct 15;96(8):2717–22.

[50] Ruggeri A, Labopin M, Sormani MP, Sanz G, Sanz J, Volt F, Michel G, Locatelli F, Diaz De Heredia C, O'Brien T, Arcese W, Iori AP, Querol S, Kogler G, Lecchi L, Pouthier F, Garnier F, Navarrete C, Baudoux E, Fernandes J, Kenzey C, Eapen M, Gluckman E, Rocha V, Saccardi R; Eurocord; Cord Blood Committee EBMT; Netcord. Engraftment kinetics and graft failure after single umbilical cord blood transplantation using a myeloablative conditioning regimen. Haematologica. 2014 Sep;99(9):1509–15.

[51] Rubinstein P, Carrier C, Scaradavou A, Kurtzberg J, Adamson J, Migliaccio AR, Berkowitz RL, Cabbad M, Dobrila NL, Taylor PE, Rosenfield RE, Stevens CE. Outcomes among 562 recipients of placental-blood transplants from unrelated donors. N Engl J Med. 1998 Nov 26;339(22):1565–77.

[52] Cohen YC, Scaradavou A, Stevens CE, Rubinstein P, Gluckman E, Rocha V, Horowitz MM, Eapen M, Nagler A, Shpall EJ, Laughlin MJ, Daniely Y, Pacheco D, Barish-

ev R, Olmer L, Freedman LS. Factors affecting mortality following myeloablative cord blood transplantation in adults: a pooled analysis of three international registries. Bone Marrow Transplant. 2011 Jan;46(1):70–6.

[53] Ooi J, Takahashi S, Tomonari A, Tsukada N, Konuma T, Kato S, Kasahara S, Sato A, Monma F, Nagamura F, Iseki T, Tojo A, Asano S. Unrelated cord blood transplantation after myeloablative conditioning in adults with acutemyelogenous leukemia. Biol Blood Marrow Transplant. 2008 Dec;14(12):1341–7.

[54] Tucunduva L, Ruggeri A, Sanz G, Furst S, Socié G, Michallet M, Arcese W, Milpied N, Yakoub-Agha I, Linkesch W, Cornelissen J, Mannone L, Iori AP, Ribera JM, Sanz J, Montesinos P, Purtill D, Labopin M, Gluckman E, Mohty M, Rocha V. Risk factors for outcomes after unrelated cord blood transplantation for adults with acute lymphoblastic leukemia: a report on behalf of Eurocord and the Acute Leukemia Working Party of the European Group for Blood and Marrow Transplantation. Bone Marrow Transplant. 2014 Jul;49(7):887–94.

[55] Brunstein C, Barker JN, Weisdorf D, DeFor TE, Miller JS, Blazar BR, McGlave PB, Wagner JE. Umbilical cord blood transplantation after nonmyeloablative conditioning: impact on transplant outcomes in 110 adults with hematological disease. Blood. 2007;110(8):3064–3070.

[56] Oran B, Wagner JE, DeFor TE, Weisdorf DJ, Brunstein CG. Effect of conditioning regimen intensity on acute myeloid leukemia outcomes after umbilical cord blood transplantation. Biol Blood Marrow Transplant. 2011 Sep;17(9):1327–34.

[57] Uchida N, Wake A, Takagi S, Yamamoto H, Kato D, Matsuhashi Y, Matsumura T, Seo S, Matsuno N, Masuoka K, Kusumi E, Yuji K, Miyakoshi S, Matsuzaki M, Yoneyama A, Taniguchi S. Umbilical cord blood transplantation after reduced-intensity conditioning for elderly patients with hematologic diseases. Biol Blood Marrow Transplant. 2008 May;14(5):583–90.

[58] Barker JN, Weisdorf DJ, DeFor TE, Blazar BR, Miller JS, Wagner JE. Rapid and complete donor chimerism in adult recipients of unrelated donor umbilical cord blood transplantation after reduced intensity conditioning. Blood. 2003;102(5): 1915–1919.

[59] Mori T, Tanaka M, Kobayashi T, Ohashi K, Fujisawa S, Yokota A, Fujita H, Nakaseko C, Sakura T, Nannya Y, Takahashi S, Kanamori H, Kanda Y, Sakamaki H, Okamoto S; Kanto Study Group for Cell Therapy. Prospective multicenter study of single-unit cord blood transplantation with myeloablative conditioning for adult patients with high-risk hematologic malignancies. Biol Blood Marrow Transplant. 2013 Mar;19(3): 486–91.

[60] MacMillan ML, Weisdorf DJ, Brunstein CG, Cao Q, DeFor TE, Verneris MR, Blazar BR, Wagner JE. Acute graft-versus-host disease after unrelated donor umbilical cord blood transplantation: analysis of risk factors. Blood. 2009 Mar 12;113(11):2410–5.

[61] Rocha V, Labopin M, Sanz G, Arcese W, Schwerdtfeger R, Bosi A, Jacobsen N, Ruutu T, de Lima M, Finke J, Frassoni F, Gluckman E; Acute Leukemia Working Party of

European Blood and Marrow Transplant Group; Eurocord-Netcord Registry. Transplants of umbilical-cord blood or bone marrow from unrelated donors in adults with acute leukemia. N Engl J Med2004;351:2276–2285.

[62] Grewal SS, Barker JN, Davies SM, Wagner JE. Unrelated donor hematopoietic cell transplantation: marrow or umbilical cord blood? Blood 2003;101:4233–4244.

[63] Gluckman E, Rocha V, Arcese W, Michel G, Sanz G, Chan KW, Takahashi TA, Ortega J, Filipovich A, Locatelli F, Asano S, Fagioli F, Vowels M, Sirvent A, Laporte JP, Tiedemann K, Amadori S, Abecassis M, Bordigoni P, Diez B, Shaw PJ, Vora A, Caniglia M, Garnier F, Ionescu I, Garcia J, Koegler G, Rebulla P, Chevret S; Eurocord Group. Factors associated with outcomes of unrelated cord blood transplant: guidelines for donor choice. Exp Hematol. 2004;32(4):397–407.

[64] Narimatsu H, Miyakoshi S, Yamaguchi T, Kami M, Matsumura T, Yuji K, Murashige N, Kusumi E, Kodama Y, Komatsu T, Sakamaki H, Kouzai Y, Okada M, Osugi Y, Kobayashi R, Inoue M, Takahashi S, Kai S, Kato K, Inoue-Nagamura T, Taniguchi S, Kato S; Japan Cord Blood Bank Network. Chronic graft-versus-host disease following umbilical cord blood transplantation: retrospective survey involving 1072 patients in Japan. Blood. 2008 Sep 15;112(6):2579–82.

[65] Newell LF, Flowers ME, Gooley TA, Milano F, Carpenter PA, Martin PJ, Delaney C. Characteristics of chronic GVHD after cord blood transplantation. Bone Marrow Transplant. 2013 Oct;48(10):1285–90.

[66] Laughlin MJ, Eapen M, Rubinstein P, Wagner JE, Zhang MJ, Champlin RE, Stevens C, Barker JN, Gale RP, Lazarus HM, Marks DI, van Rood JJ, Scaradavou A, Horowitz MM. Outcomes after transplantation of cord blood or bone marrow from unrelated donors in adults with leukemia. N Engl J Med. 2004 Nov 25;351(22):2265–75.

[67] Warlick ED, Peffault de Latour R, Shanley R, Robin M, Bejanyan N, Xhaard A, Brunstein C, Sicre de Fontbrune F, Ustun C, Weisdorf DJ, Socie G. Allogeneic hematopoietic cell transplantation outcomes in acute myeloid leukemia: similar outcomes regardless of donor type. Biol Blood Marrow Transplant. 2015 Feb;21(2):357–63.

[68] Avery S, Shi W, Lubin M, Gonzales AM, Heller G, Castro-Malaspina H, Giralt S, Kernan NA, Scaradavou A, Barker JN. Influence of infused cell dose and HLA-match on engraftment after double-unit cord blood allografts. Blood. 2011 Mar 24;117(12): 3277–85.

[69] Rocha V, Crotta A, Ruggeri A, Purtill D, Boudjedir K, Herr AL, Ionescu I, Gluckman E; Eurocord Registry. Double cord blood transplantation: extending the use of unrelated umbilical cord blood cells for patients with hematological diseases. Best Pract Res Clin Haematol. 2010Jun;23(2):223–9.

[70] Barker JN, Weisdorf DJ, DeFor TE, Blazar BR, McGlave PB, Miller JS, Verfaillie CM, Wagner JE. Transplantation of two partially HLA-matched umbilical cord blood

units to enhance engraftment in adults with hematologic malignancy. Blood 2005;105(3):1343–1347.

[71] Brunstein CG, Gutman JA, Weisdorf DJ, Woolfrey AE, Defor TE, Gooley TA, Verneris MR, Appelbaum FR, Wagner JE, Delaney C. Allogeneic hematopoietic cell transplantation for hematologic malignancy: relative risks and benefits of double umbilical cord blood. Blood. 2010;116(22):4693–4699.

[72] Somers JA, Braakman E, van der Holt B, Petersen EJ, Marijt EW, Huisman C, Sintnicolaas K, Oudshoorn M, Groenendijk-Sijnke ME, Brand A, Cornelissen JJ. Rapid induction of single donor chimerism after double umbilical cord blood transplantation preceded by reduced intensity conditioning: results of the HOVON 106 phase II study. Haematologica. 2014 Nov;99(11):1753–61.

[73] Purtill D, Smith K, Devlin S, Meagher R, Tonon J, Lubin M, Ponce DM, Giralt S, Kernan NA, Scaradavou A, Stevens CE, Barker JN. Dominant unit CD34+ cell dose predicts engraftment after double-unit cord blood transplantation and is influenced by bank practice. Biol Blood Marrow Transplant. 2013 Jun;19(6):904–11.

[74] Ponce DM, Gonzales A, Lubin M, Castro-Malaspina H, Giralt S, Goldberg JD, Hanash AM, Jakubowski A, Jenq R, Papadopoulos EB, Perales MA, van den Brink MR, Young JW, Boulad F, O'Reilly RJ, Prockop S, Small TN, Scaradavou A, Kernan NA, Stevens CE, Barker JN. Graft-versus-host disease after double-unit cord blood transplantation has unique features and an association with engrafting unit-to-recipient HLA match. Blood. 2014 Nov 6;124(19):2905–12.

[75] Ruggeri A, Sanz G, Bittencourt H, Sanz J, Rambaldi A, Volt F, Yakoub-Agha I, Ribera JM, Mannone L, Sierra J, Mohty M, Solano C, Nabhan S, Arcese W, Gluckman E, Labopin M, Rocha V; Eurocord and Acute Leukemia Working Party of European Blood and Marrow Transplant Group. Comparison of outcomes after single or double cord blood transplantation in adults with acute leukemia using different types of myeloablative conditioning regimen, a retrospective study on behalf of Eurocord and the Acute Leukemia Working Party of EBMT. Leukemia. 2014 Apr;28(4):779–86.

[76] Barker JN, Fei M, Karanes C, Horwitz M, Devine S, Kindwall-Keller TL, Holter J, Adams A, Logan B, Navarro WH, Riches M; RCI BMT 05-DCB Protocol Team. Results of a prospective multicentre myeloablative double-unit cord blood transplantation trial in adult patients with acute leukaemia and myelodysplasia. Br J Haematol. 2015 Feb;168(3):405–12.

[77] Abedin S, Peres E, Levine JE, Choi S, Yanik G, Couriel DR. Double umbilical cord blood transplantation after novel myeloablative conditioning using a regimen of fludarabine, busulfan, and total lymphoid irradiation. Biol Blood Marrow Transplant. 2014 Dec;20(12):2062–6.

[78] Chen YB, Aldridge J, Kim HT, Ballen KK, Cutler C, Kao G, Liney D, Bourdeau G, Alyea EP, Armand P, Koreth J, Ritz J, Spitzer TR, Soiffer RJ, Antin JH, Ho VT. Reduced intensity conditioning stem cell transplantation: comparison of double umbili-

cal cord blood and unrelated donor grafts. Biol Blood Marrow Transplant. 2012;18(5): 805–12.

[79] Brunstein CG, Fuchs EJ, Carter SL, Karanes C, Costa LJ, Wu J, Devine SM, Wingard JR, Aljitawi OS, Cutler CS, Jagasia MH, Ballen KK, Eapen M, O'Donnell PV; Blood and Marrow Transplant Clinical Trials Network. Alternative donor transplantation after reduced intensity conditioning: results of parallel phase 2 trials using partially HLA-mismatched related bone marrow or unrelated double umbilical cord blood grafts. Blood. 2011;118(2):282–8.

[80] Eapen M, Rocha V, Sanz G, Scaradavou A, Zhang MJ, Arcese W, Sirvent A, Champlin RE, Chao N, Gee AP, Isola L, Laughlin MJ, Marks DI, Nabhan S, Ruggeri A, Soiffer R, Horowitz MM, Gluckman E, Wagner JE; Center for International Blood and Marrow Transplant Research; Acute Leukemia Working Party Eurocord (the European Group for Blood Marrow Transplantation); National Cord Blood Program of the New York Blood Center. Effect of graft source on unrelated donor haemopoietic stem cell transplantation in adults with acute leukaemia: a retrospective analysis. Lancet Oncol. 2010;11(7):653–60.

[81] Windrum P, Morris TC, Drake MB, Niederwieser D, Ruutu T. Variation in dimethyl sulfoxide use in stem cell transplantation: a survey of EBMT centres. Bone Marrow Transplant. 2005 Oct;36(7):601–3.

[82] Morris C, de Wreede L, Scholten M, Brand R, van Biezen A, Sureda A, Dickmeiss E, Trneny M, Apperley J, Chiusolo P, van Imhoff GW, Lenhoff S, Martinelli G, Hentrich M, Pabst T, Onida F, Quinn M, Kroger N, de Witte T, Ruutu T; Chronic Malignancies and Lymphoma Working Parties of EBMT. Should the standard dimethyl sulfoxide concentration be reduced? Results of a European Group for Blood and Marrow Transplantation prospective noninterventional study on usage and side effects of dimethyl sulfoxide. Transfusion. 2014 Oct;54(10):2514–22.

[83] Abbruzzese L, Agostini F, Durante C, Toffola RT, Rupolo M, Rossi FM, Lleshi A, Zanolin S, Michieli M, Mazzucato M. Long term cryopreservation in 5% DMSO maintains unchanged CD34(+) cells viability and allows satisfactory hematological engraftment after peripheral blood stem cell transplantation. Vox Sang. 2013 Jul; 105(1):77–80.

[84] Smagur A, Mitrus I, Giebel S, Sadus-Wojciechowska M, Najda J, Kruzel T, Czerw T, Gliwinska J, Prokop M, Glowala-Kosinska M, Chwieduk A, Holowiecki J. Impact of different dimethyl sulphoxide concentrations on cell recovery, viability and clonogenic potential of cryopreserved peripheral blood hematopoietic stem and progenitor cells. Vox Sang. 2013 Apr;104(3):240–7.

[85] Keung YK, Lau S, Elkayam U, Chen SC, Douer D. Cardiac arrhythmia after infusion of cryopreserved stem cells. Bone Marrow Transplant 1994;14:363–7.

[86] Stroncek DF, Fautsch SK, Lasky LC, Hurd DD, Ramsay NK, McCullough J. Adverse reactions in patients transfused with cryopreserved marrow. Transfusion. 1991;31:521–6.

[87] Sputtek A, Jetter S, Hummel K, Kühnl P. Cryopreservation of peripheral blood progenitor cells: characteristics of suitable techniques. Beitr Infusionsther Transfusionsmed. 1997;34:79–83.

[88] Dijkstra-Tiekstra MJ, Setroikromo AC, Kraan M, Gkoumassi E, de Wildt-Eggen J. Optimization of the freezing process for hematopoietic progenitor cells: effect of precooling, initial dimethyl sulfoxide concentration, freezing program, and storage in vapor-phase or liquid nitrogen on in vitro white blood cell quality. Transfusion. 2014 Dec;54(12):3155–63.

[89] Valeri CR, Pivacek LE. Effects of the temperature, the duration of frozen storage, and the freezing container on in vitro measurements in human peripheral blood mononuclear cells. Transfusion. 1996;36:303–8

[90] Rubinstein P, Dobrila L, Rosenfield RE, Adamson JW, Migliaccio G, Migliaccio AR, Taylor PE, Stevens CE. Processing and cryopreservation of placental/umbilical cord blood for unrelated bone marrow reconstitution. Proc Natl Acad Sci U S A. 1995 Oct 24;92(22):10119–22.

[91] Harris DT, Schumacher MJ, Rychlik S, Booth A, Acevedo A, Rubinstein P, Bard J, Boyse EA. Collection, separation and cryopreservation of umbilical cord blood for use in transplantation. Bone Marrow Transplant. 1994;13:135–143

[92] Bertolini F, Corsini C, Lauri E, Gorin F, Sirchia G. J Hematother. 1993;2:233–234.

[93] Pahwa RN, Fleischer A, Than S, Good RA. Proc Natl Acad Sci U S A 1994;91:4485–4488.

[94] Rubinstein P, Carrier C, Scaradavou A, Kurtzberg J, Adamson J, Migliaccio AR, Berkowitz RL, Cabbad M, Dobrila NL, Taylor PE, Rosenfield RE, Stevens CE. Outcomes among 562 recipients of placental-blood transplants from unrelated donors. N Engl J Med. 1998 Nov 26;339(22):1565–77.

[95] Solves P, Mirabet V, Roig R. Volume reduction in routine cord blood banking. Curr Stem Cell Res Ther. 2010 Dec;5(4):362–6. Review.

[96] Solves P, Planelles D, Mirabet V, Blanquer A, Carbonell-Uberos F. Qualitative and quantitative cell recovery in umbilical cord blood processed by two automated devices in routinecord blood banking: a comparative study. Blood Transfus. 2013 Jul; 11(3):405–11.

[97] Antoniewicz-Papis J, Lachert E, Woźniak J, Janik K, Łętowska M. Methods of freezing cord blood hematopoietic stem cells. Transfusion. 2014 Jan;54(1):194–202.

[98] Yang H, Pidgorna A, Loutfy MR, Shuen P. Effects of interruptions of controlled-rate freezing on the viability of umbilical cord blood stem cells. Transfusion. 2015 Jan; 55(1):70–8.

[99] Broxmeyer HE, Srour EF, Hangoc G, Cooper S, Anderson SA, Bodine DM. High-efficiency recovery of functional hematopoietic progenitor and stem cells from human cord blood cryopreserved for 15 years. Proc Natl Acad Sci U S A. 2003 Jan 21;100(2): 645–50.

[100] Goodwin HS, Grunzinger LM, Regan DM, McCormick KA, Johnson CE, Oliver DA, Mueckl KA, Alonso JM 3rd, Wall DA. Long term cryostorage of UC blood units: ability of the integral segment to confirm both identity and hematopoietic potential. Cytotherapy. 2003;5(1):80–6.

[101] Mitchell R, Wagner JE, Brunstein CG, Cao Q, McKenna DH, Lund TC, Verneris MR. Impact of long-term cryopreservation on single umbilical cord blood transplantation outcomes. Biol Blood Marrow Transplant. 2015 Jan;21(1):50–4.

Chapter 2

Progress in Cryopreservation of Stem Cells and Immune Cells for Cytotherapy

Zhiquan Shu, Shelly Heimfeld, Zhongping Huang, Carolyn Liu and Dayong Gao

Additional information is available at the end of the chapter

http://dx.doi.org/10.5772/60620

Abstract

Cellular therapy with stem and immune cells has demonstrated significant success both in clinical treatments and the industrial market. Cryopreservation is a necessary and essential component of cellular therapy. In this chapter, first of all, some basic theories of cryoinjury and techniques in cryopreservation are reviewed. Then it focuses on the progress of cryopreservation of stem cells and immune cells, including new protocols and techniques, alternative cryoprotective agents (CPA), side effects after transplantation, and advances in reducing adverse reactions. Strategies to minimize adverse effects include medication before and after transplantation, optimizing the infusion procedure, reducing the CPA concentration or using alternative CPAs for cryopreservation, and removing CPA prior to infusion. Traditional and newly developed approaches including methods and devices for CPA removal are discussed. Future work is recommended including further optimization of cryopreservation protocols especially for lymphocytes; standardization of the optimized protocols with temperature monitoring and quality control; exploration of DMSO-free, serum-free, and even xeno-free media for cryopreservation; development of simple, reliable, and cost-effective devices for cryopreservation; and more fundamental cryobiological studies to avoid cellular injury.Keywords: cryopreservation, stem cell, immune cell, cytotherapy

Keywords: cryopreservation, stem cell, immune cell, cytotherapy

INTECH
open science | open minds

© 2015 The Author(s). Licensee InTech. This chapter is distributed under the terms of the Creative Commons Attribution License (http://creativecommons.org/licenses/by/3.0), which permits unrestricted use, distribution, and reproduction in any medium, provided the original work is properly cited.

1. Introduction

Cell therapy products, including stem cells and the stem cell-derived immune cells, with a more than $600 million market size, have been approved for use in the USA and EU[1]. With the anticipated approval of additional products, the cell therapy market is expected to grow at an annualized rate of greater than 40% through the remainder of the decade, reaching greater than $5 billion by 2020. Stem cell therapy in the USA cleared a record $250 million in 2013. All these data prove the significance of cell therapy both in clinical practice and industry.

Importantly, for most types of transplants, cryopreservation is a necessary and essential component of the clinical protocol. Long-term storage provides a solution to various logistical aspects such as the obligatory time interval between collection of the cell product, treatment with high-dose therapy, and subsequent infusion of the product in the case of autologous transplantation, or in the case of transplantation with mismatch between supply (when the baby is born) and demand (when the patient is ready to receive the unit). Cryopreservation also supports better cell therapy product characterization and quality control, improved donor screening for human leukocyte antigen (HLA) or other markers that can impact successful outcomes, and optimal transportation from the point of collection to the site of infusion.

Meanwhile, cryopreservation can be a critical issue determining the success or failure of cell transplantation. Typical cell therapeutic products must be [1] harvested from culture vessels and concentrated, followed by [2] cryopreservation permitting the cells to be stored or shipped. Cryopreservation process generally includes the addition of cryoprotective agents (CPA), cooling of samples with a specific optimal cooling rate, which is cell-type dependent, and then storage at a low temperature (such as -80°C or -196 to -150°C, temperature of liquid nitrogen or the vapor phase of liquid nitrogen). [3] Finally, the CPA must be removed post-thaw in preparation for patient administration. Herein, each step may cause injuries to cells, and therefore influence the graft function. Optimization of cryopreservation of a cell product requires understanding the mechanisms of injury caused in cryopreservation (cryoinjury), knowledge of the cryobiological characteristics of the cells, and the development of techniques and instruments to perform the optimal protocol.

In this chapter, first of all, some basic theories of cryoinjury and techniques in cryopreservation are reviewed. Then it focuses on the progress of cryopreservation in cellular therapy of two kinds of cells: stem cells and stem cell-derived immune cells.

Transplantation of stem cells, especially hematopoietic stem cells (HSC) has been successfully developed as a part of treatment protocols for a large number of clinical indications. In this part, we briefly review the progress in stem cell cryopreservation and transplantation, the side effects after transplantation along with strategies for reducing adverse reactions. Direct infusion of cryopreserved cell products into patients has been associated with the development of adverse reactions, ranging from relatively mild symptoms to much more serious, life-threatening complications. Strategies to minimize the adverse effects are reviewed. In many cases, the CPA — typically dimethyl sulfoxide (DMSO) — is believed to cause these adverse reactions and thus many studies recommend depletion of DMSO before cell infusion. Traditional and newly developed approaches including methods and devices for CPA removal are discussed.

Immune cells function to protect our body from diseases, including detection of a wide variety of pathogens, distinguishing them from the body's own healthy cells, killing infected cells, and others. They exit the bone marrow and may stay in circulation, or recirculate between blood and tissues, or go to tissues and stay there until they die. Transplantation of immune cells has been applied for the treatments of many diseases. However, despite many reports of success in immune cell cryopreservation, some conflicting and negative results are also shown in the literature. In this chapter, progresses in cryopreservation protocols and techniques for immune cells are also reviewed.

2. Fundamental cryobiology and cryopreservation processes

Generally, cryopreservation consists of several steps: preprocessing, CPA addition, cooling, storage, thawing, CPA removal, and postprocessing. Each step can affect the viability of cells after cryopreservation. Optimizing each step is necessary to maximize the retention of cell viability and functionalities. Obviously, the better understanding of the cellular injury mechanisms in cryopreservation can help minimize cellular injury. In this part, some fundamental knowledge in cryobiology is summarized. Table 1 shows the general process, cellular injury mechanisms, and proposed optimization work for cell cryopreservation [2].

Procedures	**Injury mechanisms**	**Keys to success**
Preprocessing		Cell isolation, dilution or concentration, etc.
CPA addition	Osmotic injury, CPA cytotoxicity	CPA selection (with high cryoprotective function and low cytotoxicity), CPA concentration, CPA addition (gradual addition at lower temperature).
Cooling	Mazur's "Two-Factors": ice injury and solution injury	Cooling rate and cooling history profile, reliable and simple instruments for controlled cooling, monitoring of sample temperatures during cooling.
Storage		Monitoring of sample temperatures during storage. Storage at -80°C for weeks to months and storage in liquid nitrogen or vapor phase of liquid nitrogen for months to years.
Thawing	Ice injury (recrystallization)	Fast and uniform thawing is recommended.
CPA removal	Osmotic injury, CPA cytotoxicity	Stepwise removal of CPA is recommended.
Postprocessing		Sometimes, resting the thawed cells is beneficial for the recovery of cell functions.

Table 1. Cryopreservation procedures, injury mechanisms, and optimization

2.1. Freezing of cells

Optimization of isolated cell cryopreservation requires a quantitative understanding of the biophysical response of cells during the freezing process [3-6]. As cells are cooled to between about- 5°C and -15°C, ice forms in the external medium but the cellular contents remain unfrozen and supercooled, presumably because the plasma membrane blocks the growth of ice crystals into the cytoplasm. The supercooled water inside the cells has a higher chemical potential than that of water in the partially frozen extracellular solution, so water flows out of the cell and freezes externally.

The subsequent physical phenomena in the cells depend on the cooling rate. These cell responses to freezing were first expressed quantitatively by Mazur [6] and directly linked with cell cryoinjury by Mazur's "Two-Factors" hypothesis: (a) at slow cooling rates, cryoinjury occurs due to a "solution effect", that is, the intracellular solute/electrolyte concentration increases as water leaves the cell, to a point where severe cell dehydration occurs, and (b) at high cooling rates, water is not lost fast enough and cryoinjury occurs due to intracellular ice formation (**IIF**), which ruptures the cell membrane. The optimal cooling rate for cell survival should be slow enough to reduce IIF but fast enough to minimize the solution effects. The freezing behavior of the cells can be modified by the addition of cryoprotective agents (**CPA**s), which affect the rates of water transport, ice nucleation, and ice crystal growth.

More detailed information about theoretical cryobiology can be found in the paper published in 1970 by Mazur in *Science* [5]. Important milestones in cryobiology since then include the development of cryomicroscopy allowing the observation of cell behavior during freezing and thawing [7, 8], devices to model and measure cell membrane permeabilities [9-13], and mathematical modeling to describe the probability of IIF as a function of cooling rate, temperature, and cell type[14, 15]. Karlsson and Toner incorporated the effect of CPA addition on IIF formation into these models and successfully predicted IIF formation as a function of cooling rate, temperature, and CPA concentration, leading to optimal cooling protocols preventing IIF[16].

2.2. Thawing of cells

Cells that have survived cooling to low temperatures still face the challenges of thawing, which can exert effects on survival comparable to those of cooling [5, 17]. The effects depend on whether the prior rate of cooling has induced intracellular freezing or cell dehydration. In the former case, rapid thawing can rescue many cells, possibly because it can prevent the harmful growth of small intracellular ice crystals into larger crystals by recrystallization. After severe cell dehydration, no rescue from cell death is possible.

2.3. Addition and removal of CPAs

Cells require equilibration with molar concentrations of CPAs to survive freezing. Previous research has shown a high concentration of CPA to be beneficial for preventing IIF and severe cell dehydration, thus leading to increased cell survival [18-24]. However, these CPAs have dramatic osmotic effects on cells. Cells exposed to molar concentrations of permeating CPAs undergo extensive initial dehydration followed by rehydration and potential gross swelling when the CPAs are removed. Unless precautions are taken, this shrinkage and/or swelling can be extensive enough to cause cell damage and death. Knowledge of cell membrane permeability to water and CPAs allows the prediction of the minimal and maximal cell volume excursions during addition and removal of CPAs, thus providing for a quantitative optimization approach (e.g. stepwise increase or decrease of CPA concentration in cell suspensions) to avoid osmotic damage [19-21].

2.4. Storage

Biological specimens can be stored at low temperatures for extended periods because the chemical reactions and metabolism will be slowed down or even stopped at those low temperatures. The "life clock" of these materials (RNA/DNA, cells, tissues, etc.) will be slowed or halted during cryo-storage, and then resumed after thawing to normal temperatures. The most widely applied methods for long-term storage are preserving in -80°C freezers or liquid nitrogen tanks (in liquid or vapor phases of liquid nitrogen). However, in order to totally stop any reactions, storage temperature should be below the glass transition temperature (T_g) of the aqueous solution (e.g., T_g = -123 °C for concentrated aqueous DMSO solutions [25]). Therefore, storage in liquid nitrogen tanks is recommended for preservation for months and years. Some reports have suggested that storage in the vapor phase rather than the liquid phase of liquid nitrogen would be better to avoid potential contamination between samples.

Procedures	Recommended protocol	Comments
CPA selection	· Mostly, 10% DMSO in basal medium supplemented with proteins, sometimes DMSO with lower concentrations (5–10%)	· Lower DMSO concentration leads to lower cytotoxicity. · DMSO-free, serum-free, and even xeno-free cryomedia are desired.
CPA addition	· Add CPA stock solution (precooled at 4°C, 2x concentration) to cell suspension stepwise at 4°C or room temperature (with final volume ratio 1:1). Cell concentration can be 10^7–10^8 cells/mL. · Equilibration (e.g., for 10min)	· Slow CPA addition can reduce cell volume excursion and osmotic injury. · Lower temperature may reduce CPA cytotoxicity.
Cooling	· Controlled slow cooling to a temperature (e.g., -40°C) with 1-2.5°C/min, then cooling down to a lower temperature (e.g., -80°C) at a little faster rate (about 3-5°C/min) · Put into -80°C freezer or liquid nitrogen tank.	· Slow cooling is desired to reduce ice injury to cells. · Temperature recording in whole process is recommended. · Intervention and compensation for crystallization heat by transient chilling during freezing procedure is beneficial for cell viability.
Storage	· Storage in -80°C freezer, vapor phase of liquid nitrogen, or liquid nitrogen.	· -80°C for storage for weeks to months. · Liquid nitrogen for storage for months to years.

Procedures	Recommended protocol	Comments
		· Storage in vapor phase rather than liquid phase of liquid nitrogen could decrease the contamination risk. · Temperature monitoring in whole process is recommended.
Thawing	· Fast thawing by stirring in 37°C water bath until thoroughly melted.	· Fast thawing is beneficial to reduce recrystallization and ice injury.
CPA removal	· Add washing medium slowly to the thawed cell suspension followed by equilibration and washing. · Repeated dilution/washing may be needed.	· Slow CPA removal can reduce cell volume excursion and osmotic injury. · DMSO may be associated with adverse reactions after transplantation without CPA removal; however, so far, no specific consensus exists regarding removal of DMSO prior to infusion, instead leaving the decision to physicians and clinical institutions.

Table 2. Cryopreservation protocol for HSC

3. Progress in HSC cryopreservation

3.1. HSC cryopreservation protocol

Stem cell transplantation has been performed using hematopoietic stem cells (HSC) from allogeneic, autologous, and syngeneic donors. In addition to bone marrow, HSC collected from mobilized peripheral blood or umbilical cord blood are currently in widespread clinical use, with the potential for transplantation of HSC derived from embryonic stem cell or induced pluripotent stem cell sources in the not-too-distant future [21,22]. Since the first studies of HSC freezing by Barnes and Loutit in 1955 [23], many experiments have been performed to optimize cryopreservation protocols to enhance the overall recovery and functional capacity of HSC after freezing–thawing and transfusion. Numerous excellent reviews of stem cell cryopreservation have been published, ranging from basic scientific principles to clinical cell processing protocols [24-28]. The most widely applied cryopreservation protocols for HSC have the following general features: after collection, cells are washed and resuspended in a basal saline solution supplemented with some proteins, which also contains one or more CPA. Dimethyl sulfoxide (DMSO) is the most commonly used CPA, typically at a final concentration of 5-10% (v/v). The cell suspension is frozen using a rate-controlled freezer or mechanical passive cooling methods with an optimal cooling rate of -1 to -2.5°C/min [27,28] to a subzero temperature such as -40°C [25,27,29,30], then cooled down to a lower temperature (e.g., -80°C) at a little faster rate (about 3-5°C/min), followed by transferring to a liquid nitrogen tank for long-term storage at temperatures below -150°C. Just prior to transplantation, the cryopreserved cell products are thawed quickly in a 37°C water bath and infused immediately into the patient. Sometimes, DMSO is removed before transplantation.

Table 2 shows the generally applied protocol for HSC cryopreservation. In order to maximize the quality of the frozen samples, following the optimal protocol in each step is recommended, such as gradual addition and removal of CPA, controlled freezing of samples by reliable cooling rate controlled device, fast thawing, and temperature recording during the whole process of cooling, storage, and transportation.

3.2. Adverse reactions after HSC transplantation

Infusion of thawed products has been associated with several types of adverse reactions (ARs), ranging from mild events like nausea/vomiting, hypotension or hypertension, abdominal cramps, diarrhea, flushing and chills to more severe life-threatening events like cardiac arrhythmia, encephalopathy, acute renal failure, and respiratory depression. In some cases, these adverse reactions have been directly attributed to DMSO [26-29], while others have suggested additional factors such as red cell lysate [30, 31], or infusion of high numbers of damaged granulocytes that do not survive cryopreservation [32] are the main causal trigger of these adverse reactions. To minimize such ARs after infusion, many institutions have chosen to limit the total amount of DMSO that can be infused at any one time, while others have evaluated washing protocols to first remove the DMSO and other damaged cell products prior to infusion [33-36]. In addition, patient-specific factors such as age, weight, gender, specific disease, or the type of prior treatments given and chemotherapeutic agents received can also contribute to the development of ARs, as can the infusion procedure itself (speed of injection, pausing for short periods, and the time gap between thawing of frozen cells and infusion). These topics have been intensively reviewed by Shu et al. [37], including ARs after HSC transplantation with cryopreserved products, the physiological role of DMSO in these adverse events, strategies to reduce these ARs, and new options for removal of DMSO before transfusion.

3.3. Reducing the ARs after infusion of cryopreserved HSC grafts

Many approaches have been applied to reduce the adverse effects after transplantation of cryopreserved HSC, such as:

1. systematic premedication before infusion [38],
2. hydration and allopurinol administration after infusion [38],
3. slowing the infusion speed and prolonging the infusion time [36, 38],
4. dividing the infusion into multiple aliquots given several hours or days apart [38, 39],
5. further concentrating HSC grafts to reduce the cryopreservation volumes and corresponding DMSO content [36],
6. reducing DMSO concentration for cryopreservation to lower than 10%, or using alternative CPA to mix with or replace DMSO [36, 40-42]; and
7. removing DMSO before infusion [33-36].

Since the side effects are idiosyncratic and unpredictable so far to our knowledge, all these approaches are suggested to be combined together to reduce the reaction incidence as low as possible. Several studies examining the use of lower concentrations of DMSO or alternative CPAs have been conducted, such as 7.5% DMSO+ 3% hydroxyethyl starch (HES), 5% DMSO + 6% HES, 5% DMSO+6% pentastarch [40, 42, 43]. Simply reducing DMSO concentration may decrease the toxicity and improve the kinetics of engraftment; however, it is also likely to reduce the recovery rate of HSC after cryopreservation and thawing as well. Therefore, other cryoprotective agents, such as HES or trehalose, are recommended to combine with any proposed reduction in DMSO concentration.

3.4. Removal of DMSO

A summary of options of methods and devices used for removal of DMSO from cryopreserved products is presented in Table 3 [37]. Conventional manual methods of removing DMSO from cell suspensions based on centrifugation have not changed much since the 1970s. The most widely used procedure was proposed in 1995 [44]. This process can result in cell clumping and HSC loss, cell activation, and carries a risk of product contamination. This procedure is also time-consuming and labor intensive. Several commercially developed devices have been evaluated for CPA removal, such as the CytoMate™, Sepax S-100, and Cobe-2991 instruments. Using user-definable programs DMSO can be efficiently reduced by these automated systems, resulting in reduced labor and reduced risk of contamination due to the closed fluid path. However, these devices are expensive, and again can cause cell clumping, osmotic injury, and loss of cells since they are all still based on centrifugation as their primary mode of operation.

Several new methods/technology for DMSO removal without using centrifugation have recently been developed. Fleming et al. investigated an elegant microfluidic method for small samples based on diffusion [45, 46]. However, this method has not yet been scaled up for the preparation of HSC units for transplantation. Ding et al. proposed a dialysis method for DMSO removal using hollow fiber modules with semipermeable membranes [47, 48]. Zhou and Shu et al. have developed a novel dilution-filtration method and system [49, 50], which can be used to precisely control the removal process to effectively reduce CPA concentration and prevent cell osmotic injury. Research data suggest this method promises to be a fast, safe, easy-to-operate, automated, and cost-effective approach with low cell loss and low contamination risk.

Methods or devices	Working mechanism	Comments
Centrifugation [44]	Centrifugation	Most widely applied procedure for CPA removal. · Pros: conventional devices available widely · Cons: high time and labor consumption, cell loss, high risk of contamination
CytoMate™ [51, 52]	Filtration by spinning membrane	· Pros: automated, effective and allowing a step-by-step user definable programming, low risk of contamination · Cons: high cost and cell loss due to clumping

Methods or devices	Working mechanism	Comments
Sepax S-100 [53, 54]	Consisting steps of dilution and centrifugation using a rotating syringe	· Pros: fast, automated processing, low risk of contamination · Cons: high cost and cell loss due to clumping
Cobe 2991 [35, 55]	Centrifugation	· Pros: fast, automated processing, low risk of contamination · Cons: high cost and cell loss due to clumping
Microfluidic method [45, 46]	Diffusion-based extraction in microfluidic channels	· Pros: automated processing, elegant, effective for CPA removal for samples with small volumes · Cons: hard to scale up to large volumes
Dialysis through hollow-fiber dialyzer [47, 48]	Dialysis across semipermeable hollow fiber membranes	· Pros: automated processing, effective CPA removal, low risk of contamination · Cons: high cell loss, optimization needed for samples with small volume
Dilution-filtration through hollow-fiber dialyzer [49, 50]	Controlled dilution and controlled filtration through semipermeable hollow fiber membranes	· Pros: fast, automated processing, low risk of contamination, low-cost, controllable, effective CPA removal · Cons: optimization needed for samples with small volume

Table 3. Methods and devices for CPA removal

To go along with these approaches, DMSO-washing solutions are needed. Generally, washing solutions consist of saline or cell culture medium together with nonpermeable macromolecules (dextran, albumin and/or acid citrate dextrose (ACD)), which are nontoxic, infusible, and provide a mild hyperosmotic environment to help extract the DMSO from cells. This is also why slow addition of such solutions (e.g., dripping) is preferable, as it allows the cells to slowly equilibrate to the changing osmotic environment, and minimizes the rapid uptake of water that can damage the cell membranes.

In conclusion, many progresses on effective devices and methods for removal of DMSO from cryopreserved HSC grafts have been achieved in the last decade, but challenges still remain: further studies are needed to develop the optimal (fast, safe, simple, automated, controllable, effective, and low-cost) methods and devices for CPA removal with minimum cell loss and damage.

3.5. Alternative CPAs for stem cell cryopreservation

Although DMSO has been widely accepted and utilized for stem cell cryopreservation, in some situations it may be desirable to employ other CPAs, in combination with or even replacing DMSO. The criteria for optimal CPAs include: (1) protecting cells during cryopreservation; (2) being nontoxic and able to be metabolized or digested by the body with minimal effects, which eliminates the necessity for CPA removal prior to infusion; and (3) cost and availability. Some agents, such as ethylene glycol, hydroxycellulose, sucrose, maltose, trehalose, and some macromolecules (dextran, hydroxyethyl starch, Polyvinylpyrrolidone, etc.) could be potentially used as alternative CPAs. In the last two decades, trehalose has drawn a lot of interest in this field due to its unique properties. It has very high glass transition temperature, and is extremely effective in forming a fragile glass state to protect cells during freezing/thawing and drying, maintaining the thermodynamic stability of cell membranes, and inhibiting lipid-phase transition and separation during freezing and drying [56-58]. However, for HSC, DMSO is still the most widely used CPA. In the future, searching for alternative CPAs could be another strategy to reduce the adverse reactions after HSC transplantation with DMSO.

Another possible significant development in alternative CPAs would be the development of novel, DMSO-free, serum-free, or even xeno-free cryoprotection media for the cryopreservation of stem cells. This effort will address the problems of DMSO toxicity and immunological depression posed by the presence of DMSO, animal serum, and xeno-component to the recipient after transplantation. Thirumala et al. demonstrated the feasibility of a serum-free media with reduced DMSO concentration for cryopreservation of adipose-derived adult stem cells [59]. Schulz et al. showed important advancements toward a xeno-free, chemically defined cryopreservation medium for peripheral blood mononuclear cells (PBMCs) [60,61]. These studies demonstrate the possibility of finding novel cryopreservation media for stem cells.

4. Progress in immune cell cryopreservation

4.1. Immune cell cryopreservation protocol

Cryopreservation of immune cells has been continuously studied since the 1960s. Improvement of the cryopreservation protocol was conducted based on intensive research, including fundamental cryobiological studies and experimental trials. Successful cryopreservation of lymphocytes (e.g., PBMC) has been reported and applied in clinical practice; however, negative or conflicting results also exist in the literature, especially if assessments of lymphocyte functionality rather than cell viability were applied to evaluate the impacts of cryopreservation.

Procedures	Recommended protocol	Comments
CPA selection	· Basal medium supplemented with proteins (e.g., RPMI 1640 + fetal calf serum or human serum or plasma) and DMSO (6-10% v/v)	· Lower DMSO concentration leads to lower cytotoxicity. · DMSO-free, serum-free, and even xeno-free cryomedia are desired.
CPA addition	· Add CPA stock solution (precooled at 4°C, 2x concentration) to cell suspension stepwise at 4°C with final volume ratio 1:1 and Cell concentration 10^6–10^7 cells/mL. · Equilibration (e.g., for 10 min)	· Slow CPA addition can reduce cell volume excursion and osmotic injury. · Lower temperature may reduce CPA cytotoxicity.

Procedures	Recommended protocol	Comments
Cooling	· Cooling to a temperature (e.g., -40°C) with 1°C/min by a cooling rate controlled freezer, then to a lower temperature (e.g., -80°C) at faster rate. · Put into -80°C freezer or liquid nitrogen.	· Slow cooling is desired to reduce ice injury to cells. · Temperature recording in whole process is recommended. · Intervention and compensation for crystallization heat by transient chilling during freezing (-6 to -8°C) is beneficial for cell viability.
Storage	· Storage in -80°C freezer, vapor phase of liquid nitrogen, or liquid nitrogen.	· -80°C for storage for weeks to months. · Liquid nitrogen for storage for months to years. · Storage in vapor phase, instead of in liquid phase of nitrogen, could decrease the risk of contamination/infection. · Temperature monitoring in whole process is recommended.
Thawing	· Fast thawing by stirring in 37°C water bath.	· Fast thawing is beneficial to reduce recrystallization and ice injury.
CPA removal	· Add washing medium (e.g., RPMI 1640 + fatal calf serum) stepwise at 4°C or room temperature to cell suspension followed by equilibration and centrifugation. · Repeated dilution/washing may be needed.	· Slow CPA removal can reduce cell volume excursion and osmotic injury.
Resting	· Incubation at 37°C with 5% CO_2 for a minimum of 120 min.	· Some results showed resting of cryopreserved lymphocyte can promote cell viability and functionality [62, 63]. · Incubation with cytokine cocktail may increase the retention of T cell function [64].

Table 4. Cryopreservation protocol for immune cells

Similar to the facts described above, each step in cryopreservation can affect the final status of the immune cells after freezing and thawing. Optimizing each step is necessary to minimize the potential injuries to the cells and maximize the retention of the cell functionality after processing. Table 4 shows the currently recommended protocol for immune cell cryopreservation, which is very close to that for HSC.

4.2. Impacts of cryopreservation on immune cell functions

Viability of immune cells after freezing/thawing varies from 60 to 90%. Some investigators have reported that lymphocyte functionality can be well-retained after cryopreservation. Wang et al. found that although freezing damage could result in about 10% loss of human T lymphocytes isolated from peripheral blood, cytokine producing capability was well-retained, and no apparent change in cell cycle pattern could be detected in T lymphocytes after cryostorage for 3-50 days compared to fresh samples [62]. Human peripheral lymphocytes maintained the capacity to respond to antigenic and mitogenic stimulation and to produce micronuclei in cytokinesis-blocked cells [65-67]. Cryopreservation does not induce alterations in lymphocyte surface markers and karyotypes [63, 68, 69]. Riccio et al. showed that with 24 h resting after thawing, the frequency of spontaneous apoptosis in cryopreserved lymphocyte cells was not significantly modified [70]. Hori et al. suggested that freeze–thawing did not affect the viability, phenotype, subsequent maturation, or function of dendritic cells at any stage of maturation [71].

However, alternations of lymphocyte functionality caused by cryopreservation have also been reported in the literature. Mononuclear cells (MNC) are more sensitive than stem cells. Some reports showed that cryopreservation can cause detrimental damage to lymphocyte functions [72-76]. Costantini et al. and Owen et al. found that cryopreservation could induce a consistent set of changes in PBMC from both healthy and HIV-infected patients, including a profound decrease of surface marker expression, significant changes of proportions of some cell types, loss of proliferative responses to some HIV antigens [77, 78]. Results suggested that cryopreservation induced higher levels of apoptosis in PBMC [78-80]. DNA repair capacity (DRC) has a profound influence on DNA stability and ultimately cancer incidence. Fresh human lymphocytes could repair hydrogen-peroxide-induced DNA strand breakage, while cryopreserved lymphocytes did not possess this capability [81]. Similarly, Chang found that the DRC of the cryopreserved peripheral lymphocytes was on average 14% lower than that of the fresh samples, possibility due to the destruction of DNA repair enzymes during cryopreservation [82]. A small delay in the activation of cryopreserved PBMC was also found, which implies the importance of knowing at which time points the desired cellular analyses should be carried out [83]. These results imply that special precautions should be taken in the selection of fresh or frozen lymphocyte samples for analysis and the interpretation of immune studies performed on cryopreserved lymphocytes with different functionality assays.

It is interesting that different types of immune cells, immune cells from different parts of human body (peripheral blood, lymphoid, and mucosa), or immune cells from different donors (healthy and infected patients) may have different performance in cryopreservation. Scheiwe et al. found that granulocytes were fragile and exquisitely sensitive to freezing/thawing and osmotic stress. Granulocytes could not survive cryopreservation like lymphocytes, and the damaged granulocytes might be the reason of aggregation [84]. Thus far, there is still no clinically available method to preserve granulocytes. Lymphoid dendritic cells were shown less robust than macrophages to stresses [85]. For comparison between different lymphocyte donors, more losses of T cell responses in HIV-infected individuals were detected than healthy individuals. T cells from HIV-infected donors were more fragile and more susceptible to freezing and thawing [78]. Differences were also found in the cryobiological characteristics of the immune cells. Hallak et al. demonstrated that cell membrane permeabilities to water depended on the donors' health condition and age [86]. This difference in cell membrane property may explain the different performance of those immune cells in surviving cryopreservation.

Further optimization of cryopreservation protocols for immune cells is still needed to maximize the retention of their functionality after freezing/thawing. More intensive studies of the injury mechanisms to the immune cells are desired, including in cryobiology and physiology.

Some investigators suggested a few changes to attempt to improve lymphocyte function after cryopreservation. For example, Jennes et al. found incubation of the thawed T cells with a cytokine cocktail was helpful [64]. Stroh et al. demonstrated the function of caspase inhibitors in preventing cryoinjury and improving cell recovery [80].

5. Summary

In summary, cryopreservation of stem cells and immune cells is essential for both clinical treatments and fundamental researches. Numerous studies have been done to optimize the cryopreservation protocols. The currently optimal procedure of cryopreservation of stem cells and immune cells is composed of several critical procedures: slow addition of CPA (5–10% DMSO in basal medium), slow cooling at 1–2 °C/min to a low temperature with cooling rate controlled freezer, storage in -80°C freezer or liquid nitrogen tank, fast thawing in 37°C water bath with agitation, and gradual removal of CPA prior to infusion. In the future, desired work includes further optimization of the cryopreservation protocols especially for lymphocytes; standardization of the optimized protocols with temperature monitoring and quality control; exploration of DMSO-free, serum-free or even xeno-free media for cryopreservation; development of simple, reliable, and cost-effective devices that can be used at field sites for CPA addition/removal, controlled cooling and transportation (cold chain); and more fundamental cryobiological studies to avoid injuries to the cells.

Acknowledgements

The authors are grateful to Dr. Florian Hladik and Mr. Sean M. Hughes for the enlightening discussions and their help in manuscript editing. The work was supported by NIH grants (P30 DK56465, P01 CA18029, and P30 CA15704).

Author details

Zhiquan Shu[1], Shelly Heimfeld[2], Zhongping Huang[3], Carolyn Liu[4] and Dayong Gao[1*]

*Address all correspondence to: dayong@u.washington.edu

1 Department of Mechanical Engineering, University of Washington, Seattle, WA, USA

2 Clinical Research Division, Fred Hutchinson Cancer Research Center, Seattle, WA, USA

3 Department of Biomedical Engineering, Widener University, Chester, PA, USA

4 Swedish Physicians Pine Lake Clinic, Sammamish, WA, USA

References

[1] Mason C, Manzotti E. Regenerative medicine cell therapies: numbers of units manufactured and patients treated between 1988 and 2010. *Regen Med*. 2010;5(3):307-13. Epub 2010/05/12.

[2] Shu Z, Gao D, Pu LL. Update on cryopreservation of adipose tissue and adipose- derived stem cells. *Clin Plastic Surg*. 2015;42(2):209–18. Epub 2015/04/02.

[3] Mazur P, Schmidt JJ. Interactions of cooling velocity, temperature, and warming velocity on the survival of frozen and thawed yeast. *Cryobiology*. 1968;5(1):1-17. Epub 1968/07/01.

[4] Mazur P. Equilibrium, quasi-equilibrium, and nonequilibrium freezing of mammalian embryos. *Cell Biophys*. 1990;17(1):53-92. Epub 1990/08/01.

[5] Mazur P. Cryobiology: the freezing of biological systems. *Science*. 1970;168(3934): 939-49. Epub 1970/05/22.

[6] Mazur P. Kinetics of water loss from cells at subzero temperatures and likelihood of intracellular freezing. *J Gen Physiol*. 1963;47(2):347-69.

[7] Diller KR. Quantitative low temperature optical microscopy of biological systems. J Microscopy. 1982;126(Pt 1):9-28. Epub 1982/04/01.

[8] Toner M, Cravalho EG, Karel M, Armant DR. Cryomicroscopic analysis of intracellular ice formation during freezing of mouse oocytes without cryoadditives. *Cryobiology*. 1991;28(1):55-71.

[9] Chen HH, Purtteman JJ, Heimfeld S, Folch A, Gao D. Development of a microfluidic device for determination of cell osmotic behavior and membrane transport properties. *Cryobiology*. 2007;55(3):200-9. Epub 2007/09/25.

[10] Devireddy RV, Raha D, Bischof JC. Measurement of water transport during freezing in cell suspensions using a differential scanning calorimeter. *Cryobiology*. 1998;36(2): 124-55.

[11] Gao DY, Benson CT, Liu C, McGrath JJ, Critser ES, Critser JK. Development of a novel microperfusion chamber for determination of cell membrane transport properties. *Biophys J*. 1996;71(1):443-50. Epub 1996/07/01.

[12] Gao DY, McGrath JJ, Tao J, Benson CT, Critser ES, Critser JK. Membrane transport properties of mammalian oocytes: a micropipette perfusion technique. *J Reprod Fert*. 1994;102(2):385-92. Epub 1994/11/01.

[13] Levin RL, Cravalho EG, Huggins CE. A membrane model describing the effect of temperature on the water conductivity of erythrocyte membranes at subzero temperatures. *Cryobiology*. 1976;13(4):415-29. Epub 1976/08/01.

[14] Toner M, Cravalho EG, Karel M. Thermodynamics and kinetics of intracellular ice formation during freezing of biological cells. *J Appl Phys.* 1990;67(3):1582-93.

[15] Toner M, Cravalho EG, Karel M. Cellular-response of mouse oocytes to freezing Stress - prediction of intracellular ice formation. *J Biomech Eng-T Asme.* 1993;115(2): 169-74.

[16] Karlsson JOM, Cravalho EG, Toner M. A model of diffusion-limited ice growth inside biological cells during freezing. *J Appl Phys.* 1994;75(9):4442-5.

[17] Mazur P. Freezing of living cells – mechanisms and implications. *Am J Physiol.* 1984;247(3):C125-C42.

[18] Leibo SP, Farrant J, Mazur P, Hanna MG, Jr., Smith LH. Effects of freezing on marrow stem cell suspensions: interactions of cooling and warming rates in the presence of PVP, sucrose, or glycerol. *Cryobiology.* 1970;6(4):315-32. Epub 1970/01/01.

[19] Schneider U, Mazur P. Osmotic consequences of cryoprotectant permeability and its relation to the survival of frozen-thawed embryos. *Theriogenology.* 1984;21(1):68-79.

[20] Mazur P, Schneider U. Osmotic responses of preimplantation mouse and bovine embryos and their cryobiological implications. *Cell Biophys.* 1986;8(4):259-85.

[21] Gao DY, Liu J, Liu C, Mcgann LE, Watson PF, Kleinhans FW, et al. Prevention of osmotic injury to human spermatozoa during addition and removal of glycerol. *Hum Reprod.* 1995;10(5):1109-22.

[22] Gilmore JA, Mcgann LE, Liu J, Gao DY, Peter AT, Kleinhans FW, et al. Effect of cryoprotectant solutes on water permeability of human spermatozoa. *Biol Reprod.* 1995;53(5):985-95.

[23] Fahy GM, Levy DI, Ali SE. Some emerging principles underlying the physical properties, biological actions, and utility of vitrification solutions. *Cryobiology.* 1987;24(3): 196-213.

[24] Fahy GM. The relevance of cryoprotectant toxicity to cryobiology. *Cryobiology.* 1986;23(1):1-13.

[25] Sputtek A, Kuhnl P, Rowe AW. Cryopreservation of erythrocytes, thrombocytes, and lymphocytes. *Transfus Med Hemoth.* 2007;34(4):262-7.

[26] Zambelli A, Poggi G, Da Prada G, Pedrazzoli P, Cuomo A, Miotti D, et al. Clinical toxicity of cryopreserved circulating progenitor cells infusion. *Antican Res.* 1998;18(6B):4705-8. Epub 1999/01/19.

[27] Hoyt R, Szer J, Grigg A. Neurological events associated with the infusion of cryopreserved bone marrow and/or peripheral blood progenitor cells. *Bone Marrow Transpl.* 2000;25(12):1285-7. Epub 2000/06/29.

[28] Benekli M, Anderson B, Wentling D, Bernstein S, Czuczman M, McCarthy P. Severe respiratory depression after dimethylsulphoxide-containing autologous stem cell in-

fusion in a patient with AL amyloidosis. *Bone Marrow Transpl.* 2000;25(12):1299-301. Epub 2000/06/29.

[29] Zenhausern R, Tobler A, Leoncini L, Hess OM, Ferrari P. Fatal cardiac arrhythmia after infusion of dimethyl sulfoxide-cryopreserved hematopoietic stem cells in a patient with severe primary cardiac amyloidosis and end-stage renal failure. *Ann Hematol.* 2000;79(9):523-6. Epub 2000/10/24.

[30] Salmon JP, Michaux S, Hermanne JP, Baudoux E, Gerard C, Sontag-Thull D, et al. Delayed massive immune hemolysis mediated by minor ABO incompatibility after allogeneic peripheral blood progenitor cell transplantation. *Transfusion.* 1999;39(8):824-7. Epub 1999/09/30.

[31] Sauer-Heilborn A, Kadidlo D, McCullough J. Patient care during infusion of hematopoietic progenitor cells. *Transfusion.* 2004;44(6):907-16. Epub 2004/05/26.

[32] Calmels B, Lemarie C, Esterni B, Malugani C, Charbonnier A, Coso D, et al. Occurrence and severity of adverse events after autologous hematopoietic progenitor cell infusion are related to the amount of granulocytes in the apheresis product. *Transfusion.* 2007;47(7):1268-75. Epub 2007/06/22.

[33] Bauwens D, Hantson P, Laterre PF, Michaux L, Latinne D, De Tourtchaninoff M, et al. Recurrent seizure and sustained encephalopathy associated with dimethylsulfoxide-preserved stem cell infusion. *Leuk Lymph.* 2005;46(11):1671-4.

[34] Hirata Y, Kishino K, Onozaki F, Nakaki Y, Fujiwara S, Yamamoto C, et al. Use of cryoprotectant-depleted allogeneic peripheral blood stem cells for transplantation. *Hematology.* 2011;16(4):221-4.

[35] Fois E, Desmartin M, Benhamida S, Xavier F, Vanneaux V, Rea D, et al. Recovery, viability and clinical toxicity of thawed and washed haematopoietic progenitor cells: analysis of 952 autologous peripheral blood stem cell transplantations. *Bone Marrow Transpl.* 2007;40(9):831-5. Epub 2007/08/29.

[36] Akkok CA, Holte MR, Tangen JM, Ostenstad B, Bruserud O. Hematopoietic engraftment of dimethyl sulfoxide-depleted autologous peripheral blood progenitor cells. *Transfusion.* 2009;49(2):354-61. Epub 2008/11/05.

[37] Shu Z, Heimfeld S, Gao D. Hematopoietic SCT with cryopreserved grafts: adverse reactions after transplantation and cryoprotectant removal before infusion. *Bone Marrow Transpl.* 2014;49(4):469-76.

[38] Ferrucci PF, Martinoni A, Cocorocchio E, Civelli M, Cinieri S, Cardinale D, et al. Evaluation of acute toxicities associated with autologous peripheral blood progenitor cell reinfusion in patients undergoing high-dose chemotherapy. *Bone Marrow Transpl.* 2000;25(2):173-7. Epub 2000/02/16.

[39] Junior AM, Arrais CA, Saboya R, Velasques RD, Junqueira PL, Dulley FL. Neurotoxicity associated with dimethylsulfoxide-preserved hematopoietic progenitor cell infusion. *Bone Marrow Transpl.* 2008;41(1):95-6. Epub 2007/10/16.

[40] Hayakawa J, Joyal EG, Gildner JF, Washington KN, Phang OA, Uchida N, et al. 5% Dimethyl sulfoxide (DMSO) and pentastarch improves cryopreservation of cord blood cells over 10% DMSO. *Transfusion.* 2010;50(10):2158-66.

[41] Halle P, Tournilhac O, Knopinska-Posluszny W, Kanold J, Gembara P, Boiret N, et al. Uncontrolled-rate freezing and storage at -80 degrees C, with only 3.5 percent DMSO in cryoprotective solution for 109 autologous peripheral blood progenitor cell transplantations. *Transfusion.* 2001;41(5):667-73.

[42] Rowley SD, Feng Z, Chen L, Holmberg L, Heimfeld S, MacLeod B, et al. A randomized phase III clinical trial of autologous blood stem cell transplantation comparing cryopreservation using dimethylsulfoxide vs dimethylsulfoxide with hydroxyethylstarch. *Bone Marrow Transpl.* 2003;31(11):1043-51.

[43] Donmez A, Tombuloglu M, Gungor A, Soyer N, Saydam G, Cagirgan S. Clinical side effects during peripheral blood progenitor cell infusion. *Transfus Apher Sci.* 2007;36(1):95-101. Epub 2007/01/24.

[44] Rubinstein P, Dobrila L, Rosenfield RE, Adamson JW, Migliaccio G, Migliaccio AR, et al. Processing and cryopreservation of placental/umbilical cord blood for unrelated bone marrow reconstitution. *P Natl Acad Sci USA.* 1995;92(22):10119-22. Epub 1995/10/24.

[45] Fleming KK, Longmire EK, Hubei A. Numerical characterization of diffusion-based extraction in cell-laden flow through a microfluidic channel. *J Biomech Eng-T ASME.* 2007;129(5):703-11.

[46] Fleming Glass KK, Longmire EK, Hubel A. Optimization of a Microfluidic Device for Diffusion-Based Extraction of Dmso from a Cell Suspension. *Int J Heat Mass Trans.* 2008;51(23-24):5749-57. Epub 2009/11/04.

[47] Ding WP, Zhou XM, Heimfeld S, Reems JA, Gao DY. A steady-state mass transfer model of removing CPAs from cryopreserved blood with hollow fiber modules. *J Biomech Eng-T ASME.* 2010;132(1): 011002.

[48] Ding WP, Yu JP, Woods E, Heimfeld S, Gao DY. Simulation of removing permeable cryoprotective agents from cryopreserved blood with hollow fiber modules. *J Memb Sci.* 2007;288(1-2):85-93.

[49] Shu Z, Fang C, Zhou X, Gao D. Cryoprotectant removal with dilution-filtration method and concentration monitoring with electrical conductivity measurements. *ASME* 2014 International Mechanical Engineering Congress & Exposition; November 14-20; Montreal, Canada 2014.

[50] Zhou XM, Liu Z, Shu ZQ, Ding WP, Du PA, Chung J, et al. A dilution-filtration system for removing cryoprotective agents. *J Biomech Eng-T ASME*. 2011;133(2): 021007.

[51] Calmels B, Houze P, Hengesse JC, Ducrot T, Malenfant C, Chabannon C. Preclinical evaluation of an automated closed fluid management device: cytomate (TM), for washing out DMSO from hematopoietic stem cell grafts after thawing. *Bone Marrow Transpl*. 2003;31(9):823-8.

[52] Perotti CG, Del Fante C, Viarengo G, Papa P, Rocchi L. A new automated cell washer device for thawed cord blood units. *Transfusion*. 2004;44(6):900-6.

[53] Scerpa MC, Daniele N, Landi F, Caniglia M, Cometa AM, Ciammetti C, et al. Automated washing of human progenitor cells: evaluation of apoptosis and cell necrosis. *Transf Med*. 2011;21(6):402-7.

[54] Rodriguez L, Azqueta C, Azzalin S, Garcia J, Querol S. Washing of cord blood grafts after thawing: high cell recovery using an automated and closed system. *Vox Sang*. 2004;87(3):165-72.

[55] Beaujean F, Hartmann O, Kuentz M, Le Forestier C, Divine M, Duedari N. A simple, efficient washing procedure for cryopreserved human hematopoietic stem cells prior to reinfusion. *Bone Marrow Transpl*. 1991;8(4):291-4. Epub 1991/10/01.

[56] Crowe JH, Crowe LM. Preservation of mammalian cells – learning nature's tricks. *Nat Biotechnol*. 2000;18(2):145-6.

[57] Crowe JH, Crowe LM, Chapman D. Preservation of membranes in anhydrobiotic organisms – the role of trehalose. *Science*. 1984;223(4637):701-3.

[58] Crowe JH, Crowe LM, Oliver AE, Tsvetkova N, Wolkers W, Tablin F. The trehalose myth revisited: introduction to a symposium on stabilization of cells in the dry state. *Cryobiology*. 2001;43(2):89-105.

[59] Thirumala S, Gimble JM, Devireddy RV. Evaluation of methylcellulose and dimethyl sulfoxide as the cryoprotectants in a serum-free freezing media for cryopreservation of adipose-derived adult stem cells. *Stem Cells Dev*. 2010;19(4):513-22.

[60] Germann A, Schulz JC, Kemp-Kamke B, Zimmermann H, von Briesen H. Standardized serum-free cryomedia maintain peripheral blood mononuclear cell viability, recovery, and antigen-specific T-cell response compared to fetal calf serum-based medium. *Biopreserv Biobank*. 2011;9(3):229-36. Epub 2011/10/07.

[61] Schulz JC, Germann A, Kemp-Kamke B, Mazzotta A, von Briesen H, Zimmermann H. Towards a xeno-free and fully chemically defined cryopreservation medium for maintaining viability, recovery, and antigen-specific functionality of PBMC during long-term storage. *J Immunol Methods*. 2012;382(1-2):24-31. Epub 2012/05/15.

[62] Wang SY, Hsu ML, Tzeng CH, Hsu HC, Ho CK. The influence of cryopreservation on cytokine production by human T lymphocytes. *Cryobiology*. 1998;37(1):22-9.

[63] Fujiwara S, Akiyama M, Yamakido M, Seyama T, Kobuke K, Hakoda M, et al. Cryopreservation of human-lymphocytes for assessment of lymphocyte subsets and natural-killer cytotoxicity. *J Immunol Methods.* 1986;90(2):265-73.

[64] Jennes W, Kestens L, Nixon DF, Shacklett BL. Enhanced ELISPOT detection of antigen-specific T cell responses from cryopreserved specimens with addition of both IL-7 and IL-15--the Amplispot assay. *J Immunol Methods.* 2002;270(1):99-108. Epub 2002/10/16.

[65] Zijno A, Saini F, Crebelli R. Suitability of cryopreserved isolated lymphocytes for the analysis of micronuclei with the cytokinesis-block method. *Mutagenesis.* 2007;22(5): 311-5.

[66] Thierry C, Turc JM, Valles H, Serrou B. Routine Technique of Lymphocyte Cryopreservation evaluated by in vitro tests of immune-response. *Eur J Canc.* 1977;13(4-5): 367-71.

[67] Sears HF, Tondreau SC, Rosenberg SA. Advantages of cryopreserved lymphocytes for sequential evaluation of human immune competence. II. Mixed lymphocyte cultures and mononuclear cell subpopulations. *J Nat Canc Inst.* 1978;61(4):1011-6. Epub 1978/10/01.

[68] Odonovan MR, Freemantle MR, Hull G, Arlett CF, Cole J. Extended-term culture of human T-lymphocytes – material potentially useful in genotoxicity testing. *Toxicol In Vitro.* 1994;8(4):651-3.

[69] Prince HE, Lee CD. Cryopreservation and short-term storage of human lymphocytes for surface marker analysis. Comparison of three methods. *J Immunol Methods.* 1986;93(1):15-8. Epub 1986/10/23.

[70] Riccio EKP, Neves I, Banic DM, Corte-Real S, Alecrim MD, Morgado M, et al. Cryopreservation of peripheral blood mononuclear cells does not significantly affect the levels of spontaneous apoptosis after 24-h culture. *Cryobiology.* 2002;45(2):127-34.

[71] Hori S, Heike Y, Takei M, Maruyama M, Inoue Y, Lee JJ, et al. Freeze-thawing procedures have no influence on the phenotypic and functional development of dendritic cells generated from peripheral blood CD14(+) monocytes. *J Immunother.* 2004;27(1): 27-35.

[72] Schmidtwolf IGH, Aihara M, Negrin RS, Blume KG, Chao NJ. Lymphokine-activated killer cell-activity after cryopreservation. *J Immunol Methods.* 1989;125(1-2):185-9.

[73] Kawai H, Komiyama A, Katoh M, Yabuhara A, Miyagawa Y, Akabane T. Induction of lymphokine-activated killer and natural-killer cell activities from cryopreserved lymphocytes. *Transfusion.* 1988;28(6):531-5.

[74] Lund PK, Westvik AB, Joo GB, Ovstebo R, Haug KBF, Kierulf P. Flow cytometric evaluation of apoptosis, necrosis and recovery when culturing monocytes. *J Immunol Methods.* 2001;252(1-2):45-55.

[75] Waldrop SL, Davis KA, Maine VC, Picker LJ. Normal human CD4(+) memory T cells display broad heterogeneity in their activation threshold for cytokine synthesis. *J Immunol.* 1998;161(10):5284-95.

[76] Makino M, Baba M. A cryopreservation method of human peripheral blood mononuclear cells for efficient production of dendritic cells. *Scand J Immunol.* 1997;45(6): 618-22. Epub 1997/06/01.

[77] Costantini A, Mancinia S, Giuliodoro S, Butini L, Regnery CM, Silvestri G, et al. Effects of cryopreservation on lymphocyte immunophenotype and function. *J Immunol Methods.* 2003;278(1-2):145-55.

[78] Owen RE, Sinclair E, Emu B, Heitman JW, Hirschkorn DF, Epling CL, et al. Loss of T cell responses following long-term cryopreservation. *J Immunol Methods.* 2007;326(1-2):93-115.

[79] Fowke KR, Behnke J, Hanson C, Shea K, Cosentino LM. Apoptosis: a method for evaluating the cryopreservation of whole blood and peripheral blood mononuclear cells. *J Immunol Methods.* 2000;244(1-2):139-44. Epub 2000/10/18.

[80] Stroh C, Cassens U, Samraj A, Sibrowski W, Schulze-Osthoff K, Los M. The role of caspases in cryoinjury: caspase inhibition strongly improves the recovery of cryopreserved hematopoietic and other cells. *Faseb J.* 2002;16(12):1651-3. Epub 2002/09/19.

[81] Duthie SJ, Pirie L, Jenkinson AM, Narayanan S. Cryopreserved versus freshly isolated lymphocytes in human biomonitoring: endogenous and induced DNA damage, antioxidant status and repair capability. *Mutagenesis.* 2002;17(3):211-4. Epub 2002/04/25.

[82] Chang JL, Chen G, Lampe JW, Ulrich CM. DNA damage and repair measurements from cryopreserved lymphocytes without cell culture – a reproducible assay for intervention studies. *Environ Mol Mutagen.* 2006;47(7):503-8. Epub 2006/05/05.

[83] Jeurink PV, Vissers YM, Rappard B, Savelkoul HFJ. T cell responses in fresh and cryopreserved peripheral blood mononuclear cells: kinetics of cell viability, cellular subsets, proliferation, and cytokine production. *Cryobiology.* 2008;57(2):91-103.

[84] Scheiwe MW, Pusztai-Markos Z, Essers U, Seelis R, Rau G, Korber C, et al. Cryopreservation of human lymphocytes and stem cells (CFU-c) in large units for cancer therapy--a report based on the data of more than 400 frozen units. *Cryobiology.* 1981;18(4):344-56. Epub 1981/08/01.

[85] Taylor MJ, Bank HL. Function of lymphocytes and macrophages after cryopreservation by procedures for pancreatic islets: potential for reducing tissue immunogenicity. *Cryobiology.* 1988;25(1):1-17. Epub 1988/02/01.

[86] Hallak GJ, Wilkinson JH. Membrane-permeability in normal human lymphocytes and lymphocytes from patients with chronic lymphatic leukemia. Clin Chim Acta. 1976;69(2):341-9.

Chapter 3

Use of Monoclonal Antibodies in Conditioning Regimen in Transplantation

Pervin Topcuoglu, Sinem Civriz Bozdag and Taner Demirer

Additional information is available at the end of the chapter

http://dx.doi.org/10.5772/61509

Abstract

Monoclonal antibodies (MoAbs) alone or in the combination of conventional therapies have been used in the treatment of many benign or malign diseases. In the transplantation setting, Moabs have been generally applied as a part of conditioning regimen in the aims of the prevention of graft versus host disease and/or graft failure or treatment of underlying hematologic disease. The most frequent used moAbs for this purpose are rituximab, alemtuzumab, Gemtuzumab Ozogamicin or radioimmunoconjugates. In this chapter, we discussed the role of moAbs use in the conditioning regimens of allogeneic or autologous stem cell transplantation.

Keywords: Monoclonal antibodies, Conditioning Regimen, Stem Cell Transplantation

1. Introduction

Monoclonal antibodies (MoAbs) to treat a variety of benign and malignant diseases are used alone or in combination with conventional therapies. The use of MoAbs in autologous and allogeneic hematopoietic stem cell transplant (allo-HSCT) is subject to the following conditions for:

1. In vivo purging of graft and as a part of the conditioning regimens in the autologous or allogeneic HSCTs, and/or
2. Prevention or treatment of graft versus host disease (GvHD) developed after allo-HSCT.

The goals of the use of the MoAbs for in vivo purging of graft and/or as a part of conditioning regimens in autologous or allogeneic HSCTs are to obtain tumor-free stem cells, to reduce the recurrence, and to provide the resulting increase in the efficacy of transplantation on the

INTECH
open science | open minds

© 2015 The Author(s). Licensee InTech. This chapter is distributed under the terms of the Creative Commons Attribution License (http://creativecommons.org/licenses/by/3.0), which permits unrestricted use, distribution, and reproduction in any medium, provided the original work is properly cited.

underlying disease and the cure rate. Additionally, MoAbs in allo-HSCTs prevent the graft rejection and/or reduce the frequency and severity of GvHD. More frequent used MoAbs in the transplantation are: Rituximab, Radioimmunotherapeutics (RITs), Alemtuzumab, and Gemtuzumab Ozogamisin.

2. Rituximab

Rituximab, the chimeric anti-CD20MoAb is mostly used to treat a broad variety of B-cell non-Hodgkin's lymphomas (NHL). Rituximab shows direct activity or complement-mediated cytotoxicity and antibody-dependent cytotoxicity. There are numerous studies on the use of conditioning regimens in autologous and allogeneic HSCTs settings.

The first study was reported by Flinn *et al.* [1] including 25 patients with a variety of NHL (11 follicular lymphoma, 7 mantle cell lymphoma, 5 chronic lymphocytic leukemia or small lymphocytic leukemia, 1 lymphoblastic lymphoma, and 1 marginal zone lymphoma). In this study rituximab was used for in vivo purging during the stem cell mobilization with cyclophosphamide (Cy) and also added in the myeloablative (MA) conditioning regimen including mostly Cy plus total body irradiation (TBI) and a further dose given after the engraftment. As a result, rituximab was well tolerated, engraftment was fast, and temporary neutropenia developed in the mean of 99.5 days in six patients but clinically significant infection was not reported.

Following study on the addition of rituximab for the conditioning regimen in autologous HSCT has been published by Flohr *et al.* [2]. In this phase II study, 27 patients with a variety of B-cell NHL in both the first day of the stem cell mobilization with chemotherapy and in the conditioning regimen,-10 and -3 days at the dose of 375 mg/square meter (sqm) rituximab have been used. The overall response rate has been reported as 96%. In the median follow-up of 16 months, disease free survival (DFS) and overall survival (OS) have been estimated as 77% and 95%, respectively. Another study of Khouri *et al.* [3] have evaluated the efficacy of rituximab use in the stem cell mobilization and after the transplantation in a total of 67 patients with diffuse large B-cell lymphoma (DLBCL) and follicular lymphoma (FL). Rituximab (375 mg/sqm) was infused 1 day before chemotherapy and again administered 7 days after chemotherapy at 1000 mg/sqm. In addition, rituximab has been given to the patients on the first and eighth day of 1000 mg/sqm dose following the high-dose BEAM conditioning regimen for autologous HSCT. The results were retrospectively compared with those of a historical control group (n=30 patients) receiving the same preparative regimen without rituximab. Although neutrophil engraftment in the rituximab arm was late in a statistically significant proportion (median 10 days vs.11 days, $p < 0.05$), similar incidence of infection has been shown in the patients who received rituximab compared with the control arm. In the median 20 months of follow-up, they reported that the possibilities of progression-free survival (PFS) and OS significantly prolonged in rituximab arm (PFS: 43% versus 67%, $p = 0.004$; OS: 53% versus 80%, $p = 0.002$). In a multicenter study from 10 centers associated with the Italian group, *the Gruppo ItalianoTerapie Innovative nei Linfomi*, when retrospectively compared the high-dose therapies with and without rituximab in the patients with DLBCL and FL undergoing autologous HSCT, either as salvage or as first-line therapy for high-risk presentation, rituximab was administered in four doses during the high-dose therapy immediately before peripheral blood collection to exploit the drug's in vivo purging effect, and two additional doses were usually used after autologous-HSCT. They found the similar response rate and early transplant-related mortality between rituximab (+) and (-), but the 5-year projected PFS and OS were better in those with rituximab (+) ($p<0.0001$) [4].

Hick *et al.* have also evaluated the use of rituximab in 23 patients with relapsed FL during the mobilization of stem cells for in vivo purging and during the 8th and 24th week after autologous HSCT for a four-week maintenance treatment protocol [5]. This study showed that rituximab provided permanent molecular remission in 77% of the patients associated with significantly prolonged PFS versus those with continued polymerase chain reaction PCR positivity.

Many single-arm phase II studies including small number of patients have been reported that the addition of rituximab especially reduced intensity conditioning (RIC) regimens in allogeneic HSCT settings reduced the incidence of acute or chronic GvHD and non-relapse mortality (NRM) [7-11]. In these studies, rituximab has led to an increase of serious infections due to long-term cytopenias and prolonged hypogammaglobulinemia.

Ultimately, there is no consensus regarding the dosage and scheme of rituximab use as a part of conditioning regimen for autologous and allogeneic HSCT in these studies. In addition, it is also not sufficient in randomized controlled trials demonstrating the superiority of adding rituximab. Therefore, prospective multicenter randomized trials aiming to determinate the exact role of rituximab for in vivo purging and/or as a part of conditioning regimens should be made in lymphoma patients.

3. Radioimmunotherapeutics

Radioimmunotherapeutics (RITs) uses monoclonal antibodies directed against specific tumor antigens labeled with a particle emitting radioisotope to deliver radiation directly to the tumor. This type of treatment gives a high dose of radiation to tumor tissue and protects uninvolved tissues and organs [12-13]. Labeled antibodies to the antigen over-expression in the target tissue with radioactive substances specifically bind. For this purpose beta$^-$ particles are the most frequently used: Radioactive particles connected the MoAb slowly give out its radiation and kill the other nearby cells. This is called as cross fire. They give high tissue distribution with high nucleotides in the target tissue and homogeneous energy and provide the myeloablation or affect the large tumor mass. To achieve a favorable biodistribution of a radiolabeled monoclonal antibody, an ideal antigen would be expressed homogeneously on the tumor cell surface and would lack expression on normal cells. Lacking such an antigen, methods to target lineage-specific hematopoietic antigens, such as CD20, CD33, and CD45, have been successfully developed in the autologous and allogeneic-HSCT setting.

Currently there are two RITs in clinical use for indolent NHL [14]: Yttrium-90 ibritumomab tiuxetan (^{90}Y-IT) (Zevalin) and iodine-131 tositumomab (^{131}I-T) (Bexxar). There are studies on the use of high or standard doses as a part of the conditioning regimen for the transplantation.

3.1. High-dose RITs

Studies are generally on the use of single or combined with chemotherapy in the conditioning regimen for autologous-HSCT. In the first study, Press *et al.* conducted the phase I study in 43 relapsed B-cell lymphoma patients, and the administration of anti-CD20 and anti-CD37 antibodies labeled ^{131}I-T alone in the conditioning regimen was to evaluate the toxicity and efficacy (15). The maximum tolerated dose was 27.25 Gy. However, researches have shown that patients administered more than this dose had cardio-toxic effects. In addition, the biodistribution of the antibodies in the patients with the large spleen size and a large tumor mass were emphasized not to be good in the study. The overall response rate of 95% with high-dose RIT (84% complete response and 11% partial response) and tumor response were calculated as the median of 11 months. Subsequently, the same researchers have made a phase II study with anti-CD20 labeled ^{131}I-T in 25 patients with relapse NHL [16]. In this study, they have reported that PFS was 62% and 93% of OS with the median 2-year follow-up. Similarly, Liu *et al.* found a median PFS of 42% and 68% of OS in median 42-month follow-up [17]. This was followed by similar studies regarding the use of the high-dose RITs. However, due to the gamma radiation emitted by ^{131}I-T most subsequent studies had been conducted with 90Y-IT, a pure beta emitter [18-21]. There was no need to prolong strict patient isolation and contact alert in ^{90}Y-IT in contrast to^{131}I-T. Besides, disease statuses prior to the transplantations in those studies were also variable. Although the use of high-dose RIT was planned for the patients unable to tolerate high-dose treatment, the majority of patients in the studies was under 60 years of age and had *chemosensitive* relapses. Additionally, there are no prospective studies to prove RITs' superiority to chemotherapy and/or radiotherapy. Furthermore, this treatment should be administered in specialized centers.

3.2. Standard-dose RITs

To overcome the problems related to the safe yield of high-dose RITs, the efficacies of standard-dose RITs combined with chemotherapy in preparative regimens of the transplantation have been assessed in the following studies. The results in several single-arm studies not including control group were impressive. In a randomized trial, the Blood and Marrow Transplant Clinical Trials Clinical Trials Network (BMT-CTN) 0401 in which ^{131}I-T-BEAM or Rituximab-BEAM were given to the patients with *chemosensitive* relapse of DLBCL [22], disease control and survival effects of RIT could not have been shown. A randomized study compared ^{90}Y-IT-BEAM with BEAM alone in recurrent B cell NHL was closed early by reason of the slow patient recruitment. As their evaluation with a small number of cases, it was the first randomized study that proved that higher DFS and OS were in the RIT arm. Nevertheless, the published studies do not support the routine use of standard-dose RITs in DLBCL.

Some studies in *chemorefractory* DLBCL patients who underwent autologous-HSCT conditioned by standard-dose RIT have been reported as two or three year PFS and OS 39%–63%

and 55%–67%, respectively [23-26]. In the European Mantle Cell Network MCL-3 study ^{90}Y-IT was given to patients younger than 66 years one week prior to BEAM or BEAC (Carmustine, Etoposide, Cytarabine and Cyclophosphamide) conditioning regimen [27]. When compared with the results of the MCL-2 study, they concluded that there was no benefit in the patients undergoing autologous-HSCTas a first-line intensification treatment.

In allogeneic HSCT, RIT has generally been added to the RIC regimens in the refractory NHL patients. One of the first studies where Shimoni *et al.* gave ^{90}Y-IT with fludarabine-based conditioning regimens to 12 patients with *chemorefractory* CD20+ NHL demonstrated the safety of RIT [28]. Subsequently, several studies related to adding RITs to the conditioning regimen have been published [29,30]. Although allogeneic HSCTs made by adding RITs to RIC regimes have reliability in these studies, it has not been shown to be superior to the transplantations with RIC regimens not including RIT yet.

There are some studies related to the adding of RITs to the preparative regimens in acute leukemia and myelodysplastic syndrome (MDS) as well. Initially, ^{131}I-labeled M195, the mouse Moabs of lintuzumab (reactive with CD33) was used in 10 patients with relapsed or refractory acute myeloid leukemia in a phase I study from the Memorial Sloan Kettering Cancer Center [31]. Subsequently, ^{131}I-labeled anti-CD33 MoAbs were added to standard myeloablative regimen in 31 patients with refractory or relapsed AML (n=16), accelerated or blastic phase chronic myeloid leukemia (CML) (n=14) or advanced myelodysplastic syndrome (MDS) (n=1) underwent allogeneic HSCT from their related donor [32]. The median survival was calculated as 4.9 months (range 0.3–90+ months). Three patients with relapsed AML remain in complete remission more than 5 years.

Based on the feasibility of MoAbs, investigators have focused on the CD45, the other antigen. The CD45 antigen, common leukocyte antigen, is expressed on the surface of virtually all hematopoietic cells, except mature red cells and platelets. In addition, CD45 expression has been detected in 85% to 90% of AML and ALL, and the antigen does not internalize after the antibody binding. In a phase I study conducted by the Fred Hutchinson Cancer Center, RIT with ^{131}I-anti-CD45 has been implanted one week prior to the conditioning regimen including Cy-TBI in AML beyond the first complete remission (CR), acute lymphoblastic leukemia, and MDS-excess blast [33]. The patients of this study have undergone allogeneic HSCT from their human leukocyte antigen (HLA) -identical relative donors or autologous HSCT. This first study has shown that the radiation with acceptable toxicity should be given to the bone marrow and spleen. Subsequently RIT with ^{131}I-anti-CD45 has been added to the conditioning regimen with busulfan (Bu) plus Cy (BuCy) in the patients with the first CR AML [34]. Three-year non-relapse mortality and disease-free survival in this study was calculated at 21% and 61%, respectively. They have reported that the results were comparable by the International Bone Marrow Transplantation Registry (IBMTR) data using only BuCy in allogeneic HSCT. Similarly Pagel *et al.* added the RIT to the RIC regimens in elderly patients with advanced stage or high risk AML and showed the reliability of RIT in a phase I study [35]. Same researches used the ^{131}I-anti CD45 targeted radiotherapy in combination with Fludarabine plus 2 Gy TBI in younger patients (age 16 to 50 years) with advanced AML or high-risk MDS who underwent allogeneic HSCT [36]. They aimed to define the maximum tolerated dose (MTD) of radiolabeled anti-CD45 antibody in addition to non-myeloablative conditioning regimes and to create better antitumor control with minimal toxicity in comparison with standard myeloablative regimens. Their study suggested that a maximum dose of 28 Gy could be delivered to the liver and the arbitrary limit of 43 Gy to the marrow might be unnecessarily conservative. Conjugation of anti-CD45 antibody with alternative radioisotopes including ^{90}Y is currently explored in clinical trials.

Another attempt in the studies was the use of anti-CD66 moAbs in leukemic patients. But leukemic blasts do not express CD66. Therefore, the anti-leukemic effect of CD66 RIT depends on "crossfire" from the beta-particles emitted by 188-Rhenium (Re). ^{188}Re-labeled anti-CD66 moAbs were used as a part of the standard myeloablative conditioning regimen including total body irradiation (12 Gy) (n = 30) or busulfan (n = 27) and high-dose cyclophosphamide +/- thiotepa prior to allogeneic or autologous HSCT in 57 patients with high-risk AML or MDS [37]. In median 26 months follow-ups, disease-free survival were 64% for 44 patients in the first or second CR or in very good partial remission (less than 15% blasts in the marrow at transplantation) and only 8% for those with more than 15% blasts in the marrow at transplantation. Likewise, targeted marrow irradiation with ^{188}Re-anti-CD66 moAbs were used in 20 patients with Philadelphia chromosome-positive acute lymphoblastic leukemia or advanced CML prior to allogeneic HSCT [38]. With a median follow-up of 54 months (range 23–81) overall and disease-free survival were 29% (95%-CI 14–58) and 25% (95%-CI 12–53), respectively. Subsequently, conjugation of anti-CD66 with ^{188}Re or ^{90}Y were added to a reduced intensity conditioning regimen in 20 patients with a median age of 63 years (range: 55–65 years) suffering from acute leukemia (n=17) or MDS (n=3) [39]. The probability of survival was estimated as 70% at 1 year and 52% at 2 years post-transplant. They concluded that ^{90}Y-anti-CD66 moAbs were more feasible and less nephrotoxic than ^{188}Re.

Briefly, the use of RIT is an attractive approach to increase conditioning prior to HSCT. The randomized studies in refractory aggressive or indolent NHL show the superiority of adding RIT. Nevertheless, the addition of standard dose RIT to the conditioning regimen in autologous transplantation is a valuable research topic. In allogeneic transplantation, until displaying the superiority of RIT-based conditioning regimen in controlled randomized studies, this approach should only be considered within clinical trials.

4. Alemtuzumab (Campath)

Alemtuzumab is a human originated MoAbs to CD52 that normally expresses on B and T lymphocytes, macrophages, monocytes, natural killer cells, and some dendritic cells. While alemtuzumab efficiently reduces both T and B cells from the circulating blood, it has minimal or no effect on hematopoietic progenitor cells [40].

Anti-CD52 is often used in the treatment of chronic lymphocytic leukemia. But adding CD52 to the conditioning regimens in allogeneic HSCTs in many malign hematological diseases has reduced the frequency and severity of GvHD as well as decreased the risk of graft rejection [41-43]. Also alemtuzumab in combination with fludarabine and Cy in allogeneic HSCT for

http://dx.doi.org/10.5772/61509

acquired aplastic anemia was associated with a very low incidence of chronic GvHD and excellent survival [44-49]. However, the studies have reported that alemtuzumab led to increase the frequency of opportunistic infection, in particular Cytomegalovirus, Epstein–Barr virus, and Adenovirus, and the risk for the recurrence of the underlying disease due to the reduction of graft versus tumor effects.

5. Gemtuzumab Ozogamisin

Gemtuzumab ozogamisin (GO) is a moAbs to CD33 conjugated with human calicheamicins. It has been withdrawn from the market by the US Food and Drug Administration in 2010 because of the increasing risk of liver sinusoidal obstruction, and a lack of data for the efficacy and safety. Recently many studies have been published about the use of GO in the treatment of CD33+ AML patients as a part of induction therapy or consolidation [50-51]. Furthermore, phase I/II studies have reported that the use of GO as a part of MA or RI conditioning regimens in allo-HSCT setting could be safe and efficient in poor-risk AML patients [52-54]. In addition, a pilot study has been recently published about the administration of GO combined with azacytidine as the maintenance treatment of post-transplant relapses in AML [55].

6. Conclusion

Although many studies have been published for the additions of MoAbs to the conditioning regimens for HSCTs, there are no sufficient data to determine the optimal dose and administration schedule of the MoAbs until now. However, rituximab recently has been widely used in many single-arm studies in NHL patients who underwent allogeneic HSCT. Another controversial issue is about the use of RITs. Randomized data do not support incorporating RITs into the conditioning regimens for either autologous or allogeneic SCT settings. The high-dose RITs should be used in refractory or advanced malign hematological disease in specialized centers though lacking randomized data. The standard-dose RIT is also a good research topic for lymphoid malignancies planning high dose therapy with autologous rescue or allogeneic SCT.

Moreover, given the reduced risk of graft failure and GvHD with alemtuzumab but increased risk of disease relapse and the incidence of opportunistic infections, the use of alemtuzumab in the allogeneic SCT should be considered in patients with matched unrelated or mismatched related donors. Owing to the shortage of studies on GO, another MoAbs, with the reuse, GO should be used in clinical trials.

In conclusion, it is anticipated that additional MoAbs to the conditioning regimens will be routinely used in the next door following by the proven clinical efficacy and safety.

Author details

Pervin Topcuoglu*, Sinem Civriz Bozdag and Taner Demirer

*Address all correspondence to: topcuogl@medicine.ankara.edu.tr

Ankara University, School of Medicine, Dept of Hematology, Stem Cell Transplantation Unit, Ankara, Turkey

References

[1] Flinn IW, O'Donnell PV, Goodrich A, Vogelsand G, Abrams R, Noda S, Marcelus D, Borowitz M, Jones R, Ambinde RF. Immunotherapy with Rituximab during peripheral blood stem cell transplantation for non-Hodgkin's lymphoma. Biology of Blood and Marrow Transplantation. 2000; 6(6):628-632.

[2] Flohr T, Hess G, Kolbe K, Gamm H, Nolte H, Stanislawski T, Huber C, Derigs HG. Rituximab in vivo purging is safe and effective in combination with CD34-positive selected autologous stem cell transplantation for salvage therapy in B-NHL. Bone Marrow Transplant. 2002; 29(9):769-775.

[3] Khouri IF, Saliba RM, Hosing C, Okoroji GJ, Acholonu S, Anderlini P, Couriel D, De Lima M, Donato ML, Fayad L, Giralt S, Jones R, Korbling M, Maadani F, Manning JT, Pro B, Shpall E, Younes A, McLaughlin P, Champlin RE. Concurrent administration of high-dose rituximab before and after autologous stem-cell transplantation for relapsed aggressive B-cell non-Hodgkin's lymphomas. J ClinOncol. 2005; 23(10): 2240-2247.

[4] Tarella C, Zanni M, Magni M, Benedetti F, Patti C, Barbui T, Pileri A, Boccadoro M, Ciceri F, Gallamini A, Cortelazzo S, Majolino I, Mirto S, Corradini P, Passera R, Pizzolo G, Gianni AM, Rambaldi A. Rituximab improves the efficacy of high-dose chemotherapy with autograft for high-risk follicular and diffuse large B-cell lymphoma: a multicenter Gruppo Italiano Terapie Innovative nei linfomi survey. J ClinOncol. 2008; 26(19):3166-3175.

[5] Hicks LK, Woods A, Buckstein R, Mangel J, Pennell N, Zhang L, Imrie K, Spaner D, Cheung MC, Boudreau A, Reis M, Crump M, Berinstein NL. Rituximab purging and maintenance combined with auto-SCT: long-term molecular remissions and prolonged hypogammaglobulinemia in relapsed follicular lymphoma. Bone Marrow Transplant. 2009; 43(9):701-708.

[6] Kharfan-Dabaja MA, Pidala J, Kumar A, Terasawa T, Djulbegovic B. Comparing efficacy of reduced-toxicity allogeneic hematopoietic cell transplantation with conventional chemo-(immuno) therapy in patients with relapsed or refractory CLL: a Markov decision analysis. Bone Marrow Transplant. 2012; 47(9):1164-1170.

[7] Michallet M, Socié G, Mohty M, Sobh M, Bay JO, Morisset S, Labussière-Wallet H, Tabrizi R, Milpied N, Bordigoni P, El-Cheikh J, Blaise D. Rituximab, fludarabine, and total body irradiation as conditioning regimen before allogeneic hematopoietic stem cell transplantation for advanced chronic lymphocytic leukemia: long-term prospective multicenter study. Exp Hematol. 2013; 41(2): 127-133.

[8] Pidala J, Roman-Diaz J, Kim J, Nishihori T, Perkins J, Tate C, Ochoa-Bayona JL, Field T, Fernandez HF, Tomblyn M, Ayala E, Anasetti C, Kharfan-Dabaja MA. Targeted IV busulfanand fludarabine followed by post-allogeneic hematopoietic cell transplantation rituximab demonstrate encouraging activity in CD20+ lymphoid malignancies without increased risk of infectious complications. Int J Hematol. 2011; 93(2): 206-212.

[9] McIver Z, Stephens N, Grim A, Barrett AJ. Rituximab administration within 6 months of T cell depleted allogeneic SCT is associated with prolonged life-threatening cytopenias. Biol Blood Marrow Transplant. 2010;16(11):1549-1556.

[10] Khouri IF, Wei W, Korbling M, Turturro F, Ahmed S, Alousi A, Anderlini P, Ciurea S, Jabbour E, Oran B, Popat UR, Rondon G, Bassett RL Jr, Gulbis A. BFR (bendamustine, fludarabine, and rituximab) allogeneic conditioning for chronic lymphocytic leukemia/lymphoma: reduced myelosuppression and GVHD. Blood. 2014; 124(14): 2306-2312.

[11] Shamriz O, Vilk SR, Wolf DG, Ta-Shma A, Averbuch D, Weintraub M, Stepensky P. Hematopoietic stem cell transplantation conditioning with use of rituximab in EBV related lymphoproliferative disorders. Clin Immunol. 2014;151(2): 79-83.

[12] Kraeber-Bodéré F, Bodet-Milin C, Rousseau C, Eugène T, Pallardy A, Frampas E, Carlier T, Ferrer L, Gaschet J, Davodeau F, Gestin JF, Faivre-Chauvet A, Barbet J, Chére lM. Radioimmunoconjugates for the treatment of cancer. Semin Oncol. 2014; 41(5): 613-622.

[13] Kawashima H. Radioimmunotherapy: a specific treatment protocol for cancer by cytotoxic radioisotopes conjugated to antibodies. Scientific World Journal. 2014; 2014: 492061.

[14] Kharfan-Dabaja MA, Nishihori T, Otrock ZK, Haidar N, Mohty M, Hamadani M. Monoclonal antibodies in conditioning regimens for hematopoietic cell transplantation. Biol Blood Marrow Transplant. 2013 Sep; 19(9):1288-1300.

[15] Press OW, Eary JF, Appelbaum FR, Martin PJ, Badger CC, Nelp WB, Glenn S, Butchko G, Fisher D, Porter B, Matthews DC, Fisher LD, Bernstein ID. Radiolabeled-antibody therapy of B-cell lymphoma with autologous bone marrow support. N Engl J Med. 1993; 329(17):1219 -1224.

[16] Press OW, Eary JF, Appelbaum FR, Martin PJ, Nelp WB, Glenn S, Fisher DR, Porter B, Matthews DC, Gooley T. Phase II trial of 131I-B1 (anti-CD20) antibody therapy with autologous stem cell transplantation forrelapsed B cell lymphomas. Lancet. 1995; 346(8971): 336-340.

[17] Liu SY, Eary JF, Petersdorf SH, Martin PJ, Maloney DG, Appelbaum FR, Matthews DC, Bush SA, Durack LD, Fisher DR, Gooley TA, Bernstein ID, Press OW. Follow-up of relapsed B-cell lymphoma patients treated with iodine-131-labeled anti-CD20 antibody and autologous stem-cell rescue. J Clin Oncol. 1998; 16(10): 3270-3278.

[18] Nademanee A, Forman S, Molina A, Fung H, Smith D, Dagis A, Kwok C, Yamauchi D, Anderson AL, Falk P, Krishnan A, Kirschbaum M, Kogut N, Nakamura R, O'donnell M, Parker P, Popplewell L, Pullarkat V, Rodriguez R, Sahebi F, Smith E, Snyder D, Stein A, Spielberger R, Zain J, White C, Raubitschek A. A phase 1/2 trial of high-dose yttrium-90-ibritumomab tiuxetan in combination with high-dose etoposide and cyclophosphamide followed by autologous stem cell transplantation in patients with poor-risk or relapsed non-Hodgkin lymphoma. Blood.2005; 106(8): 2896-2902.

[19] Winter JN, Inwards DJ, Spies S, Wiseman G, Patton D, Erwin W, Rademaker AW, Weitner BB, Williams SF, Tallman MS, Micallef I, Mehta J, Singhal S, Evens AM, Zimmer M, Molina A, White CA, Gordon LI. Yttrium-90 ibritumomab tiuxetan doses calculated to deliver up to 15 Gy to critical organs may be safely combined with high-dose BEAM and autologous transplantation in relapsed or refractory B-cell non-Hodgkin's lymphoma. J ClinOncol. 2009; 27(10):1653-1659.

[20] Devizzi L, Guidetti A, Tarella C, Magni M, Matteucci P, Seregni E, Chiesa C, Bombardieri E, Di Nicola M, Carlo-Stella C, Gianni AM. High-dose yttrium-90 ibritumomab tiuxetan with tandem stem-cell reinfusion: an outpatient preparative regimen for autologous hematopoietic celltransplantation. J ClinOncol. 2008; 26(32): 5175-5182.

[21] Ferrucci PF, Vanazzi A, Grana CM, Cremonesi M, Bartolomei M, Chinol M, Ferrari M, Radice D, Papi S, Martinelli G, Paganelli G. High-activity 90Y-ibritumomab tiuxetan (Zevalin) with peripheral blood progenitor cell support in patients with refractory/resistant B-cell non-Hodgkin lymphomas. Br J Haematol. 2007; 139(4): 590-599.

[22] Vose JM, Carter S, Burns LJ, Ayala E, Press OW, Moskowitz CH, Stadtmauer EA, Mineshi S, Ambinder R, Fenske T, Horowitz M, Fisher R, Tomblyn M. Phase III randomized study of rituximab/carmustine, etoposide, cytarabine, and melphalan (BEAM) compared with iodine-131 tositumomab/BEAM with autologous hematopoietic cell transplantation for relapsed diffuse large B-cell lymphoma: results from the BMT CTN 0401 Trial. J Clin Oncol. 2013; 31(13):1662-1668.

[23] Vose JM, Bierman PJ, Loberiza FR, Enke C, Hankins J, Bociek RG, Chan WC, Weisenburger DD, Armitage JO. Phase II trial of 131iodine tositumomab with high-dose chemotherapy and autologous stem cell transplantation for relapsed diffuse large B cell lymphoma. Biol Blood Marrow Transplant. 2013; 19(1):123-128.

[24] Vose JM, Bierman PJ, Enke C, Hankins J, Bociek G, Lynch JC, Armitage JO. Phase I trial of iodine-131 tositumomab with high-dose chemotherapy and autologous stem-

cell transplantation for relapsed non-Hodgkin's lymphoma. J ClinOncol. 2005; 23(3): 461-467.

[25] Shimoni A, Zwas ST, Oksman Y, Hardan I, Shem-Tov N, Yerushalmi R, Avigdor A, Ben-Bassat I, Nagler A. Yttrium-90-ibritumomab tiuxetan(Zevalin) combined with high-dose BEAM chemotherapy and autologous stem cell transplantation for chemo-refractory aggressive non-Hodgkin's lymphoma. Exp Hematol. 2007; 35(4): 534-540.

[26] Briones J, Novelli S, Garcia Marco JA, Tomas JF, Bernal T, Grande C, Canales M, Torres A, Moraleda JM, Panizo C, Palmero F, Jarque I, Hernandez M, Barca EG, Lopez D, Caballero D. Autologous stem cell transplantation with yttrium-90 ibritumomab tiuxetan (Zevalin) plus BEAM conditioning in patients with refractory non-Hodgkin diffuse large B-cell lymphoma: results of a prospective, multicenter, Phase II clinical trial. ASH Annual Meeting Abstracts. 2012; 120:1978.

[27] Geisler CH, Kolstad A, Laurell A, Andersen NS, Pedersen LB, Jerkeman M, Eriksson M, Nordström M, Kimby E, Boesen AM, Kuittinen O, Lauritzsen GF, Nilsson-Ehle H, Ralfkiaer E, Akerman M, Ehinger M, Sundström C, Langholm R, Delabie J, Karjalainen-Lindsberg ML, Brown P, Elonen E. Nordic lymphoma group. Long-term progression-free survival of mantle cell lymphoma after intensive front-line immunochemotherapy with in vivo-purged stem cell rescue: a nonrandomized phase 2 multicenter study by the Nordic Lymphoma Group. Blood. 2008; 112(7): 2687-2693.

[28] Shimoni A, Zwas ST, Oksman Y, Hardan I, Shem-Tov N, Rand A, Yerushalmi R, Avigdor A, Ben-Bassat I, Nagler A. Ibritumomab tiuxetan (Zevalin) combined with reduced-intensity conditioning and allogeneic stem-cell transplantation (SCT) in patients with chemorefractory non-Hodgkin's lymphoma. Bone Marrow Transplant. 2008; 41(4): 355-361.

[29] Bethge WA, Lange T, Meisner C, Von Harsdorf S, Bornhaeuser M, Federmann B, Stadler M, Uharek L, Stelljes M, Knop S, Wulf G, Trenschel R, Vucinic V, Dittmann H, Faul C, Vogel W, Kanz L, Bunjes D. Radioimmunotherapy with yttrium-90-ibritumomab tiuxetan as part of a reduced-intensity conditioning regimen for allogeneic hematopoietic cell transplantation in patients with advanced non-Hodgkin lymphoma: results of a phase 2 study. Blood. 2010; 116(10):1795-1802.

[30] Khouri IF, Saliba RM, Erwin WD, Samuels BI, Korbling M, Medeiros LJ, Valverde R, Alousi AM, Anderlini P, Bashir Q, Ciurea S, Gulbis AM, de Lima M, Hosing C, Kebriaei P, Popat UR, Fowler N, Neelapu SS, Samaniego F, Champlin RE, Macapinlac HA. Nonmyeloablative allogeneic transplantation with or without 90 yttrium ibritumomab tiuxetan is potentially curative for relapsed follicular lymphoma: 12-year results. Blood. 2012; 119(26): 6373-6378.

[31] Scheinberg DA, Lovett D, Divgi CR, Graham MC, Berman E, Pentlow K, Feirt N, Finn RD, Clarkson BD, Gee TS. A Phase I trial of monoclonal antibody M195 in acute myelogenous leukemia: specific bone marrow targeting and internalization of radionuclide. J Clin Oncol. 1991; 9(3): 478-490.

[32] Burke JM1, Caron PC, Papadopoulos EB, Divgi CR, Sgouros G, Panageas KS, Finn RD, Larson SM, O'Reilly RJ, Scheinberg DA, Jurcic JG. Cytoreduction with iodine-131-anti-CD33 antibodies before bone marrow transplantation for advanced myeloid leukemias. Bone Marrow Transplant. 2003; 32(6): 549-556.

[33] Matthews DC, Appelbaum FR, Eary JF, Fisher DR, Durack LD, Hui TE, Martin PJ, Mitchell D, Press OW, Storb R, Bernstein ID. Phase I study of 131I-anti-CD45 antibody plus cyclophosphamide and total body irradiation for advanced acute leukemia and myelodysplastic syndrome. Blood. 1999; 94(4): 1237-1247.

[34] Horowitz MM, Press OW, Eary JF, Rajendran J, Fisher DR, Gooley T, Ruffner K, Nemecek E, Sickle E, Durack L, Carreras J, Horowitz MM, Press OW, Gopal AK, Martin PJ, Bernstein ID, Matthews DC. 131I-anti-CD45 antibody plus busulfan and cyclophosphamide before allogeneic hematopoietic cell transplantation for treatment of acute myeloid leukemia in first remission. Blood. 2006; 107(5): 2184-2191.

[35] Pagel JM, Gooley TA, Rajendran J, Fisher DR, Wilson WA, Sandmaier BM, Matthews DC, Deeg HJ, Gopal AK, Martin PJ, Storb RF, Press OW, Appelbaum FR.. Allogeneic hematopoietic cell transplantation after conditioning with 131 I-anti-CD45 antibody plus fludarabine and low-dose total body irradiation for elderly patients with advanced acute myeloid leukemia or high-risk myelodysplastic Syndrome. Blood. 2009; 114(27): 5444-5453.

[36] Mawad R, Gooley TA, Rajendran jG, Fisher DR, Gopal AK, Shileds AT, Sandmaier BM, Sorro ML, Deeg HJ, Strob R, Green DJ, Maloney DG, Appelbaum FR. Radioalabeled anti-CD45 Antibody with reduced-intensity conditioning and allogeneic transplantation for younger patients with advanced acute myeloid leukemia or myelodysplastic syndrome. Bio Bone Marrow Transplantation. 2014; 20: 1363-1368.

[37] Bunjes D. 188Re-labeled anti-CD66 monoclonal antibody in stem cell transplantation for patients with high-risk acute myeloid leukemia. Leuk Lymphoma. 2002; 43(11): 2125-2131.

[38] Zenz T, Glatting G, Schlenk RF, Buchmann I, Döhner H, Reske SN, Bunjes D. Targeted marrow irradiation with radioactively labeled anti-CD66 monoclonal antibody prior to allogeneic stem cell transplantation for patients with leukemia: results of a phase I-II study. Haematologica. 2006; 91(2): 285-286.

[39] Ringhoffer M, Blumstein N, Neumaier B, Glatting G, von Harsdorf S, Buchmann I, Wiesneth M, Kotzerke J, Zenz T, Buck AK, Schauwecker P, Stilgenbauer S, Döhner H, Reske SN, Bunjes D. 188Re or 90Y-labelled anti-CD66 antibody as part of a dose-reduced conditioning regimen for patients with acute leukaemia or myelodysplastic syndrome over the age of 55: results of a phase I-II study. Br J Haematol. 2005; 130(4): 604-613.

[40] Gribben JG, Hallek M. Rediscovering alemtuzumab: current and emerging therapeutic roles. Br J Haematol. 2009; 144 (6):818-831.

[41] Pérez-Simón JA, Kottaridis PD, Martino R, Craddock C, Caballero D, Chopra R, García-Conde J, Milligan DW, Schey S, Urbano-Ispizua A, Parker A, Leon A, Yong K, Sureda A, Hunter A, Sierra J, Goldstone AH, Linch DC, San Miguel JF, Mackinnon S; Spanish and United Kingdom Collaborative Groups for Nonmyeloablative Transplantation. Nonmyeloablative transplantation with or without alemtuzumab: comparison between 2 prospective studies in patients with lymphoproliferative disorders. Blood. 2002; 100(9):3121-3127.

[42] Hale G, Zhang MJ, Bunjes D, Spence D, Horowitz MM, Barrett AJ, Waldmann H. Improving the outcome of bone marrow transplantation by using CD52 monoclonal antibodies to prevent graft-versus-host disease and graft rejection. Blood. 1998; 92(12): 4581-4590.

[43] Delgado J, Thomson K, Russell N, Ewing J, Stewart W, Cook G, Devereux S, Lovell R, Chopra R, Marks DI, Mackinnon S, Milligan DW. British Society of Blood and Marrow Transplantation. Results of alemtuzumab-based reduced-intensity allogeneic transplantation for chronic lymphocytic leukemia: a British Society of Blood and Marrow Transplantation Study. Blood. 2006; 107(4):1724-1730.

[44] Peggs KS, Sureda A, Qian W, Caballero D, Hunter A, Urbano-Ispizua A, Cavet J, Ribera JM, Parker A, Canales M, Mahendra P, Garcia-Conde J, Milligan D, Sanz G, Thomson K, Arranz R, Goldstone AH, Alvarez I, Linch DC, Sierra J, Mackinnon S. UK and Spanish Collaborative Groups. Reduced-intensity conditioning for allogeneic haematopoietic stem cell transplantation in relapsed and refractory Hodgkin lymphoma: impact of alemtuzumab and donor lymphocyte infusions on long-term outcomes. Br J Haematol. 2007; 139(1):70-80.

[45] Gandhi S, Kulasekararaj AG, Mufti GJ, Marsh JC. Allogeneic stem cell transplantation using alemtuzumab-containing regimens in severe aplastic anemia. Int J Hematol. 2013; 97(5): 573-580.

[46] March JC, Gupta V, Lim Z, Ho AY, Ireland RM, Hayden J, Potter V, Koh MB, Islam MS, Russel N, Marks DI, Mufti GJ and Pagliuca A. Alemtuzumab with fludarabine and cyclophosphamide reduces Chronic graft-versus-host disease after allogeneic stem cell transplantation for acquired aplastic anemia. Blood. 2011; 118)(8): 2351-2357

[47] De Lima M, Champlin RE, Thall PF, Wang X, Martin TG 3rd, Cook JD, McCormick G, Qazilbash M, Kebriaei P, Couriel D, Shpall EJ, Khouri I, Anderlini P, Hosing C, Chan KW, Andersson BS, Patah PA, Caldera Z, Jabbour E, Giralt S. Phase I/II study of gemtuzumab ozogamicin added to fludarabine, melphalan and allogeneic hematopoietic stem cell transplantation for high-risk CD33-positive myeloid leukemias and myelodysplastic syndrome. Leukemia. 2008; 22 (2):258-264.

[48] Marsh JC, Pearce RM, Koh MB, Lim Z, Pagliuca A, Mufti GJ, Perry J, Snowden JA, Vora AJ, Wynn RT, Russell N, Gibson B, Gilleece M, Milligan D, Veys P, Samarasinghe S, McMullin M, Kirkland K, Cook G. British Society for Blood and Marrow Transplantation, Clinical Trials Committee. Retrospective study of

alemtuzumab vs ATG-based conditioning without irradiation for unrelated and matched sibling donor transplants in acquired severe aplastic anemia: a study from the British Society for Blood and Marrow Transplantation. Bone Marrow Transplant. 2014 Jan; 49(1):42-48.

[49] Hamad N, Del Bel R, Messner HA, Kim D, Kuruvilla J, Lipton JH, Uhm J, Seftel M, Alam N, Xu W, Gupta V. Outcomes of hematopoietic cell transplantation in adult patients with acquired aplastic anemia using intermediate-dose alemtuzumab-based conditioning. Biol Blood Marrow Transplant. 2014; 20(11):1722-1728.

[50] Castaigne S, Pautas C, Terre C, Raffoux E, Bordessoule D, Bastie JN, Legrand O, Thomas X, Turlure P, Reman O, De Revel T, Gastaud L, de Gunzburg N, Contentin N, Henry E, Marolleau JP, Aljijakli A, Rousselot P, Fenaux P, Preudhomme C, Chevret S, Dombret H; Acute Leukemia French Association. Effect of gemtuzumab ozogamicin on survival of adult patients with de-novo acute myeloid leukaemia(ALFA-0701): a randomised, open-label, phase 3 study. Lancet. 2012; 379(9825):1508-1516.

[51] [51]Petersdorf SH, Kopecky KJ, Slovak M, Willman C, Nevill T, Brandwein J, Larson RA, Erba HP, Stiff PJ, Stuart RK, Walter RB, Tallman MS, Stenke L, Appelbaum FR. A phase 3 study of gemtuzumab ozogamicin during induction and postconsolidation therapy in younger patients with acute myeloid leukemia. Blood 2013; 121(24): 4854–4860.

[52] Kharfan-Dabaja MA, Hamadani M, Reljic T, Pyngolil R, Komrokji RS, Lancet JE, Fernandez HF, Djulbegovic B, Kumar A. Gemtuzumab ozogamicin for treatment of newly diagnosed acute myeloid leukaemia: a systematic review and meta-analysis. Br J Haematol 2013; 163(3): 315-325.

[53] Bornhäuser M, Illmer T, Oelschlaegel U, Schetelig J, Ordemann R, Schaich M, Hänel M, Schuler U, Thiede C, Kiani A, Platzbecker U, Ehninger G. gemtuzumab ozogamicin as part of reduced-intensity conditioning for allogeneic hematopoietic cell transplantation in patients with relapsed acute myeloid leukemia. Clin Cancer Res. 2008; 14 (17):5585-5593.

[54] Satwani P, Bhatia M, Garvin JHJr, George D, Dela Cruz F, Le Gall J, Jin Z, Schwartz J, Duffy D, Van de Ven C, Foley S, Hawks R, Morris E, Baxter-Lowe LA, Cairo MS. A Phase I study of gemtuzumabozogamicin (GO) in combination with busulfan and cyclophosphamide (Bu/Cy) and allogeneic stem cell transplantation in children with poor-risk CD33(+) AML: a new targeted immunochemotherapy myeloablative conditioning (MAC) regimen. Biol Blood Marrow Transplant. 2012; 18(2):324-329.

[55] Oshikawa G1, Kakihana K, Saito M, Aoki J, Najima Y, Kobayashi T, Doki N, Sakamaki H, Ohashi K. Post-transplant maintenance therapy with azacitidine and gemtuzumab ozogamicin for high-risk acute myeloid leukaemia. Br J Haematol. 2014 Dec 17. doi: 10.1111/bjh.13248 [Epub ahead of print].

Chapter 4

Progress in Haploidentical Hematopoietic Stem Cell Transplantation

Stefan O. Ciurea and Ulas D. Bayraktar

Additional information is available at the end of the chapter

http://dx.doi.org/10.5772/61883

Abstract

Haploidentical hematopoietic cell transplantation (HaploHCT), with cells from HLA-half-matched first degree related donors (siblings, children and parents), could revolutionize hematopoietic transplantation as it expands this form of treatment to approximately 40% of patients who do not have an HLA-matched donor in USA. This need is particularly acute in developing countries, which usually do not have an unrelated donor registry and/or cost is a major issue in acquiring unrelated donor stem cells. Accordingly, the number of haploSCTs done in USA, Europe, China, and developing countries is on the rise. Advantages to HaploHCT include almost universal (more than 95% of patients will have a half-matched related donor) and immediate availability of donor progenitor cells, the opportunity to select the best donor among family members to minimize treatment-related mortality, decrease relapse rate and improve outcomes [2], and the possibility to collect donor cells for cellular therapy post-transplantation, with the goal to enhance the anti-tumor effects of the graft. Despite its potential advantages, until recently, high donor-recipient HLA-histoincompatibility has proven very difficult to overcome.

Haploidentical transplants initially performed with conventional GVHD prophylaxis in late 70's led to a strong bidirectional alloreactivity, manifested by both high incidence of primary graft failure of approximately 30% as well as the development of a syndrome suggestive of hyperacute GVHD (manifested with seizures, renal failure, respiratory failure in the majority of patients) and very poor outcomes. To prevent GVHD after HaploHCT, *ex vivo* T-cell depletion (TCD) was used successfully in the 80's [5]; however, this approach resulted in a high incidence of graft rejection in up to 50% of cases [6]. This high incidence of graft failure, thought to be primarily related to the remaining T cells in the recipients system and lack of donor T cells in the graft to support engraftment, was improved in the 90's by intensifying the conditioning regimens, combining *ex vivo* and *in vivo* T-cell depletion, and increasing the donor graft inoculum using "megadoses" of CD34+ cells. Primary engraftment was achieved in >90% patients with a low GVHD rate [8].

In the past decade, significant progress has been made as investigators from around the world have tried to overcome the fore-mentioned barriers in HaploHCT by using T-cell

INTECH
open science | open minds

© 2015 The Author(s). Licensee InTech. This chapter is distributed under the terms of the Creative Commons Attribution License (http://creativecommons.org/licenses/by/3.0), which permits unrestricted use, distribution, and reproduction in any medium, provided the original work is properly cited.

replete grafts with intensified GVHD prophylaxis, or by the use of methods to selectively deplete T cells from the haploidentical graft [12]. In addition, the development of post-transplant cellular therapy to prevent or treat disease relapse and infectious complications after transplant has found an ideal applicability in related donor transplantation, including haploidentical transplants.

Keywords: Haploidentical Hematopoietic, Stem Cell Transplantation

---

1. Introduction

A human leukocyte antigen (HLA)-matched sibling is the preferred donor for allogeneic hematopoietic stem cell transplantation (AlloSCT); however, the probability of having such a donor depends on the number of one's siblings and is approximately 30% in the population. The probability of finding a matched unrelated donor (MUD), the second preferred donor, primarily depends on a patient's race and ethnicity. While a Caucasian's chance of having a MUD is 75%, that of an African American is less than 20% [1, 2]. For those without an HLA-matched donor, alternative hematopoietic progenitor cell sources include mismatched unrelated donors, haploidentical related donors, and umbilical cord blood.

While a patient's children and parents share one haplotype with the patient, the chance that one's sibling and cousin would share at least one haplotype is 75% and 37.5%, respectively. Consequently, almost all patients with known parents have at least one haploidentical donor. The use of haploidentical donors as an alternative to HLA-matched sibling donors (MSD) has been gaining momentum recently [3], particularly after the advent of posttransplantation cyclophosphamide [4], which rendered this form of transplantation easier and safer.

The primary challenge in AlloSCT from haploidentical related donors (HaploSCT) is overcoming the high HLA histoincompatibility barrier. In fact, first HaploSCT attempts in the late 1970s led to a strong bidirectional alloreactivity, leading to both high incidence of graft failure and the development of hyperacute graft-versus-host disease (GVHD) [5, 6]. To prevent GVHD after HaploSCT, ex vivo T-cell depletion (TCD) was used in the 1980s [7]; however, this led to a high incidence of graft rejection due to the lack of T cells in the graft that would have eliminated the remaining recipient T cells [8]. Outcomes after HaploSCT were improved in the 1990s by intensifying the conditioning regimen, combining *ex vivo* and *in vivo* T-cell depletion, and increasing the donor graft inoculum using "mega-doses" of CD34+ cells [9]. This led to intensified work on HaploSCT and, in the 2000s, Johns Hopkins group introduced a forgotten method, high-dose cyclophosphamide early after graft infusion, bringing HaploSCT to mainstream use. These strategies have been improved upon and newer ones are being developed to tackle the high-HLA histoincompatibility barrier, while at the same time fastening posttransplant immune reconstitution and preserving graft-versus-leukemia (GVL) effect. These transplant approaches may be grouped into two: those including ex vivo T-cell depletion or manipulation of the graft and those relying on modification/intensification of GVHD prophylaxis without graft engineering. The strategies covered in this chapter and their rationales are summarized in Table 1.

Approach	Mechanism and rationale	Potential shortcomings
Complete/partial ex vivo T-cell depletion	Most efficacious GVHD preventive method	↑ graft rejection ↑ NRM and possible ↑ RI due to delayed immune reconstitution
T_{reg} and T_{con} co-infusion	Addition of T_{cons} to promote immune reconstitution while preventing GVHD with T_{regs}	T_{reg} may decrease GVL effect T_{reg}/T_{con} ratio needs to be optimized
NK-cell co-infusion	Addition of NK cells to enhance GVL effect and decrease TRM	Clinical efficacy not proven
Engineered donor lymphocytes with a safety suicide switch	To prevent/treat disease relapse and improve immune reconstitution post transplant. Safety switch allowing T-cell suicide in case of GVHD precipitation → higher T-cell doses are possible	T cells are not targeted → while immune reconstitutive effect is demonstrated, GVL effect is not yet clear
T cells with chimeric antigen receptors	T cells engineered to recognize specific antigens (CD19) provide GVL effect without GVHD	Clinical efficacy after HaploSCT has not been shown yet
Allodepletion using anti-CD25 antibodies	ex vivo depletion of alloreactive T cells by targeting activation marker CD25 after incubation with APC recipients	T_{reg} also express CD25 Clinical efficacy not proven Possible effect on GVL response
Allodepletion with phototoxic dye	ex vivo depletion of alloreactive T cells with TH9402 that accumulates in activated T cells	Clinical efficacy not proven Possible effect on GVL response
Selective $\alpha\beta$ T-cell depletion	Preservation of $\gamma\delta$ T cells (unlikely to induce GVHD while effective against infections with an innate-like response) while eliminating $\alpha\beta$ T cells most responsive for aGVHD Potential to avoid posttransplant immunosuppression	Clinical efficacy not proven Promising early data available Possible effect on GVL response
Selective $CD45RA^{+}$ T-cell depletion	Elimination of $CD45RA^{+}$ naïve T cells (capable of precipitating GVHD) while preserving memory T cells (active against infections) Potential to avoid posttransplant immunosuppression	Clinical efficacy not proven Possible effect on GVL response

Approach	Mechanism and rationale	Potential shortcomings
Alloanergization	Alloreactive T cells are anergized by blocking co-stimulatory CD80/86 signal	T cells are not depleted ↑ GVHD rate
High-dose posttransplantation cyclophosphamide RIC/NMA conditioning	Eliminating the alloactivated T cells early after transplant without affecting stem cells. T-cell preservation allows lower intensity conditioning extending transplantation to elderly patients Low incidence of cGVHD	Low cost GVHD incidence higher than after ex vivo T-cell depletion; however similar to matched transplantation Higher leukemia relapse incidence after NMA conditioning
Myeloablative conditioning	To decrease relapse incidence in leukemia patients	↑ in NRM and possibly in GvHD
Peripheral blood as stem cell source	To decrease relapse incidence and possibly improve immune reconstitution through higher T-cell content in PB	↑ in acute GvHD potential
Intensified immune suppression	To demeliorate immune reaction both ways G-CSF priming of BM and PB graft to induce T-cell hyporesponsiveness	Higher aGVHD and cGVHD incidence

Legend: GVHD – graft-versus-host disease, NRM – nonrelapse mortality, RI – relapse incidence, T_{reg} – regulatory T cells, T_{con} – conventional T cells, APCs – antigen-presenting cells, GVL – graft-versus-leukemia effect, RIC – reduced-intensity conditioning, NMA – nonmyeloablative conditioning, HaploHCT – haploidentical transplantation, PB – peripheral blood, BM – bone marrow, G-CSF – granulocyte–colony-stimulating factor, NK – natural killer

Table 1. The rationale and potential shortcomings of the current approaches in haploidentical stem cell transplantation.

2. HaploSCT with *ex vivo* T-Cell depletion or manipulation

The first successful HaploSCT strategy was grafting a "mega-dose" of progenitor cells through TCD of the bone marrow (BM) and peripheral blood (PB) products. To further decrease graft rejection, *in vivo* TCD with antithymocyte globulin (ATG) and a myeloablative-conditioning regimen were used [9]. Further technical revisions in the protocol led to primary engraftment in 95% of patients [10]. Although GVHD rates were low, transplant-related mortality (TRM) rate approached 40% primarily due to opportunistic viral infections, likely related to the delayed immune reconstitution. Furthermore, the use of myeloablative conditioning restricted this type of transplant to younger patients with good performance status. Two general approaches were used to enhance GVL and immune reconstitution after TCD HaploSCT: (1)

selective lymphocyte add-back during or after TCD graft infusion and (2) selective depletion or deactivation of T cells capable of inducing GVHD while preserving the rest.

2.1. Selective lymphocyte add-back during or after TCD graft infusion

2.1.1. Co-infusion of regulatory T cells (T_{reg}) and conventional T cells (T_{con})

T_{regs} modulate the immune system maintaining tolerance to self-antigens. Studies showed that T_{regs} may suppress GVHD [11] and facilitate posttransplant immune reconstitution when coinfused with T_{cons} [12]. Whether T_{regs} affect GVL effect is under investigation [13]. To boost the GVL effect and immune reconstitution with T_{cons} while preventing GVHD with T_{regs}, Di Ianni et al. infused donor T_{regs} before the infusion of TCD PB progenitor cells and donor T_{cons} [14]. Of 28 patients, 26 achieved engraftment and 2 developed acute GVHD (aGVHD). Despite the rapid development of a wide T-cell repertoire, 8 patients still died of opportunistic infections. A recent follow-up study also demonstrated high engraftment, low GVHD, and high TRM rates [15]. These findings suggest that adoptive immunotherapy with T_{regs} may counteract the GVHD potential of conventional T cells in HaploSCT; however, the high incidence of opportunistic infections and TRM remains a concern.

2.2. Natural Killer (NK) cells

It is thought that NK cells, a vital part of the innate immune system [16], recognize their targets through both inhibitory and activating receptors. According to the widely used "missing self" model, an NK cell recognizes a cell as foreign when the particular cell lacks one or more HLA class I alleles specific to the inhibitory receptors (killer immunoglobulin-like receptors, KIRs) on the NK cell [17, 18]. NK cells primarily attack hematopoietic cells sparing the solid organs; therefore, they are almost incapable of causing GVHD [19]. NK-cell infusions after HaploSCT have been utilized to exploit innate immunity against a variety of tumors [20]– [22]. Yoon et al. reported no acute side effects in 14 patients who were infused with donor NK cells 6–7 weeks after T-cell replete HaploHCT using a reduced-intensity conditioning [23]. Two patients who received NK-cell infusions during active leukemia did not have a response and 4 patients developed cGVHD. More recently, the same group reported no acute toxicity after NK-cell infusions up to 1×10^8 cells/kg. When compared with historical controls, NK-cell infusions were associated with lower leukemia relapse rate [24]. Further studies are needed to assess the utility of NK-cell infusions after HaploSCT, and such a study is currently recruiting patients at our institution.

2.3. Engineered donor lymphocytes with a safety switch

Donor lymphocyte infusions (DLI) are more practical after HaploSCT than after transplants from unrelated donors due to the availability of the related donors. While DLI may be used to prevent or treat disease relapse and enhance posttransplant immune reconstitution, it may also induce GVHD. T cells engineered to express safety suicide switches in case of GVHD may be used for a safer DLI. Ciceri et al. engineered donor lymphocytes to express *Herpes simplex* virus-thymidine kinase suicide gene (TK cells), which can be triggered by the use of ganciclovir [25]. TK cells were engrafted in 22 of 28 patients who underwent HaploSCT with TCD–PB grafts and received TK cells once a month for 4 months. Immune responses against cytomegalovirus (CMV) and Epstein–Barr virus improved after TK-cell infusions. Without any GVHD prophylaxis, 10 patients developed acute GVHD and required ganciclovir resulting in abrogation of GVHD in all. There were no GVHD-related deaths or long-term complications [25].

However, ganciclovir, a commonly used drug to treat CMV after transplantation, is not a well-suited drug to induce suicide of T cells. Baylor group used an alternative approach and engineered donor lymphocytes to express an inducible caspase-9 transgene (iC9), activated by a bio-inert molecule, AP1903 [26]. All of 10 pediatric patients (age 3–17) who underwent HaploSCT with TCD grafts and were infused with iC9-T cells between 30 and 90 days after transplantation, achieved engraftment of iC9-T cells [27]. In 5 patients who developed GVHD after iC9-T-cell infusion, iC9-T cells were >90% eliminated within 2 hours of AP1903 administration, and GVHD was rapidly reversed. Viral replication or disease was resolved within 4 weeks of iC9-T-cell infusion in all patients who had evidence of viral replication. Although very promising with a strong rationale, engineering T lymphocytes requires good manufacturing practice (GMP) facilities and patient-specific tailoring and is expensive.

2.4. T cells with Chimeric Antigen Receptors (CAR)

Lymphocytes, irrespective of whether they have been engineered to express suicide genes or not, have a broad target range that may or may not include the underlying malignancy. To give them a specific target, T cells are engineered to express CARs (CAR T cells) – fusion proteins with an extracellular antigen recognition moiety and an intracellular T-cell activation domain. CAR T cells have significantly higher antitumor efficacy for B-cell hematological malignancies without the added risk for the development of GVHD. Kochenderfer et al. reported their findings in 10 patients who received anti-CD19 CAR T cells for B-cell malignancies relapsed after transplantation from matched related or unrelated donors [28]. All patients had received standard DLIs prior to CAR T cells with only 2 responses. Two patients achieved responses lasting >3 and >9 months after CAR T-cell infusions, whereas 6 patients achieved stable disease lasting between 1 and more than 11 months. None of the patients developed GVHD after the infusion.

CAR T cells after HaploHCT may also be generated from the same donor and used to prevent/treat relapses. At our institution, we have so far treated 3 acute lymphoblastic leukemia (ALL) patients – one with active disease – with HaploSCT followed by CAR T cells. All patients tolerated the infusions well with no significant GVHD. The two patients who received CAR T cells as preemptive therapy are alive in remission more than 6 months post transplant, whereas the other patient died of disease relapse. To our knowledge, these are the first HaploSCT patients treated with CAR T cells. Although the experience is limited, the prevention of disease relapse post transplant for high-risk ALL patients appears to be the most important therapeutic benefit of CAR T cells presently.

http://dx.doi.org/10.5772/61883

2.5. Selective T-Cell depletion/deactivation of the graft

2.5.1. Allodepletion

Allodepletion methods include, first, generating an alloresponse by co-culture of donor T cells and recipient cells and, then, depleting the activated donor T cells through surface activation markers or photoactive dyes, which are preferentially retained in activated T cells [29].

Amrolia et al. used an anti-CD25 immunotoxin to deplete alloreactive lymphocytes ex vivo. Allodepleted lymphocytes of 10^4–10^5 cells/kg were infused on days 30, 60, and 90 of TCD HaploSCT in 16 patients (median age 9 years) [30]. Only two patients developed grade II–IV acute GVHD, and a wider T-cell receptor (TCR) repertoire was observed 4 months after the transplant compared with the retrospective controls. Nevertheless, 9 patients (56%) died due to relapsed disease (5), infection (3), and interstitial pneumonitis (1).

Depletion based on CD25 expression may not be the optimal approach as T_{regs} also express CD25 on their surface. An alternate method to deplete activated T cells using TH9402, a phototoxic dye that accumulates in activated T cells due to their inability to efflux rhodamide-like drugs, was also developed [29, 31, 32]. Bastien et al. showed that photodepletion in transplanted patients with resistant chronic GVHD eradicated proliferating T cells while sparing T_{regs} [33]. HaploSCT with photodepleted T cells may be possible and requires further clinical studies.

Although allodepletion has a strong rationale, clinical studies to date are limited, and its broad use is severely hampered by the requirement of cell cultures in a GMP facility.

2.5.2. Alloanergization

For activation of T cells, two signals from antigen-presenting cells (APCs) are required: presentation of the immunogenic peptide on major histocompatibility complex activating the TCR and a costimulatory or an inhibitory signal through CD80/86 and CTLA-4 on APCs, respectively, to the CD28 on T cells. Although a costimulatory signal would lead to differentiation to T_{cons}, an inhibitory signal from CTLA-4 would induce anergy and the development of T_{regs} [34] allowing transplantation of histoincompatible allografts [35].

Guinan et al. showed the feasibility of HaploSCT using a BM graft in which donor T cells were anergized through incubation with recipient's mononuclear cells and CTLA-4-Ig [36]. Of 12 patients transplanted, 1 died early post transplant, 11 achieved sustained engraftment, and 3 developed acute GVHD. No deaths due to GVHD occurred in this group. In a follow-up study, 5 of 24 transplanted patients were reported to develop severe aGVHD and 12 patients died within 200 days of transplantation (5 due to infection) [37]. Similar to allodepletion methods, use of alloanergization is restricted to those centers with GMP facilities.

2.5.3. CD45RA depletion

Various classification schemes of T cells exist according to their cell surface phenotype and functional activity [38]– [40]. Majority of T cells that can respond to minor H antigens and cause

GVHD are thought to be never exposed to their cognate antigen, in other words, naïve (T_N), with a $CD45RA^+CD62L^+$ surface phenotype [41]. Several in vitro and mouse studies support this hypothesis [42]– [46]. However, depletion of $CD45RA^+$ naïve T cells is not straightforward, as a subset of CD34+ hematopoietic progenitor cells also express CD45RA [47]. To preserve the progenitor cells, Bleakley et al. devised a two-step procedure in which donor-apheresed PB is first selected for CD34+ cells, and then CD34-negative fraction was depleted for CD45RA to preserve all CD34+ cell subsets [48]. Conversely, investigators at St. Jude chose to deplete CD3+ cells from the first day – preserving all CD34+ cells – and CD45RA+ cells from the second-day apheresis products [49]. A 4.5-log depletion in T_N cells was detected in the final product to be infused. In 8 pediatric patients who underwent HaploSCT after myeloablative conditioning, the use of CD3/CD45RA depleted grafts led to engraftment in all and development of GVHD in none of the patients [49]. On posttransplant day 30, almost all T cells were negative for CD45RA. After a median follow-up of 171 days, none of the patients died of infectious complications. Although very promising, these results need to be verified in larger cohorts and in adults.

2.5.4. Alpha-beta T-cell depletion

γδ T cells, with TCRs made up of one γ (gamma) and one δ (delta) chain, possess properties of both innate and adaptive immune system with rapid, innate-like responses and rearranged TCRs yielding adaptability [50]. Remarkably, γδ T cells are thought not to require antigen processing and HLA presentation of antigens, rendering them unlikely to induce GVHD, whereas αβ T cells are thought to be the primary cause of GVHD [51]. Accordingly, a faster recovery of γδ T cells after SCT has been associated with longer disease-free survival [52]. Recently, methods to deplete αβ T cells preserving γδ T cells have been developed [53].

Of the few clinical studies available, Bertaina et al. reported primary engraftment in 44 of 45 children (median age of 10 years) with acute leukemia who underwent HaploSCT with TCR-αβ and CD19-depleted PB grafts [54]. With the only pharmacologic GVHD prophylaxis of pre-transplant ATG, none of the patients developed grade III–IV acute GVHD, whereas 13 children developed grade I–II skin-only GVHD. Two patients died of infectious complications. After a median follow-up of 11 months, the 2-year leukemia-free survival was 75%. On a follow-up study of 23 children with nonmalignant disorders, the same strategy led to a TRM of 9.3% and grade III–IV acute GVHD was not found. As with CD45RA depletion, these results are promising but need to be verified in larger cohorts and in adults.

3. HaploSCT without graft engineering

A highly effective GVHD prevention is necessary to overcome the intense bidirectional alloreactivity (in both graft-versus-host and host-versus-graft directions) associated with HaploSCT. Ex vivo TCD is the most efficacious method to prevent GVHD; however, (1) it requires myeloablative conditioning to ensure engraftment compensating for the lack of donor T cells eradicating residual recipient immune cells, (2) it requires a relatively sophisticated cell processing laboratory, and (3) it is associated with slower recovery of cell-mediated immune system. To overcome these hurdles, either T-cell depletion methods were modified or augmented as outlined above or a robust GVHD prophylaxis regime was used in place of ex vivo TCD. The latter is typically achieved by either posttransplantation cyclophosphamide or intensification of the traditional GVHD prophylaxis.

3.1. Posttransplantation high-dose cyclophosphamide (Post-Cy)

In 1960s, Barenbaum et al. demonstrated that Post-Cy could prevent skin graft rejection when administered 2–3 days after allografting in a mouse model [55]. This forgotten method was revived by the Johns Hopkins group in the late 1990s when they showed that Post-Cy attenuated lethal and nonlethal GVHD in mice and prolonged their survival [4]. Cyclophosphamide is thought to prevent GVHD by eliminating rapidly dividing donor T cells induced by the major HLA mismatch early after the haploidentical graft infusion. Furthermore, quiescent progenitor cells and memory T cells in the graft are less susceptible to cyclophosphamide due to their high levels of aldehyde dehydrogenase [4, 56].

Post-Cy has been adapted to HaploSCT using nonmyeloablative conditioning and BM grafts that have a lower T-cell content than PB grafts [4, 57]. After various single-center reports, the multicenter BMT CTN 0603 trial demonstrated the feasibility of Post-Cy in HaploSCT with an acceptable incidence of GVHD (32% acute grade II–IV and 13% chronic GVHD) and very low TRM [58]. The disappointingly high relapse incidence (45%) was primarily attributed to the use of nonmyeloablative conditioning for patients with acute leukemias. Conversely, Post-Cy has yielded particularly impressive results in patients with lymphoma. A retrospective analysis of 151 consecutive patients with poor risk or advanced lymphoma who underwent HaploSCT with Post-Cy revealed a progression-free survival of 40% at 3 years [59], similar to what has been observed in patients with Hodgkin's disease after HLA-matched transplants [60].

3.1.1. Post-Cy after myeloablative conditioning

Relatively high relapse rates with Post-Cy approach in patients with acute leukemia prompted researchers to intensify the conditioning regimen. Early results from the Johns Hopkins group with Post-Cy after myeloablative conditioning demonstrated acceptable GVHD and engraftment rates, albeit in a pediatric and young adult cohort [61]. More recently, Raiola et al. reported grade II–III acute GVHD incidence of 12% and disease-free-survival of 68% after a median follow-up of 333 days in a cohort of 50 patients with high-risk hematological malignancies who underwent HaploSCT with Post-Cy and busulfan or total-body irradiation (TBI)-based myeloablative conditioning [62]. Our experience with Post-Cy approach using a myeloablative yet reduced-intensity conditioning with fludarabine, melphalan +/− thiotepa (subsequently changed to 2 Gy TBI) has been very good, with TRM and progression-free survival of 21% and 53% after a median follow-up of 14 months in 57 patients with advanced hematological malignancies [63]. Updated results for our first 100 patients treated showed a 3-year PFS of 56% for patients with acute myeloid leukemia (AML) in CR1/CR2 or chronic-phase CML (chronic myeloid leukemia), 62% for patients with lymphoid malignancies, and

44% for patients with advanced acute lymphoblastic leukemia [63], results comparable with matched transplants.

3.1.2. Post-Cy with peripheral blood grafts

With a higher T-cell content, the use of PB grafts may lead to faster posttransplant immune recovery and improve graft-versus-leukemia effect with the expense of higher GVHD incidence. Raj et al. reported that while the incidence of grade II–IV aGVHD appeared to be twice as much as with a BM graft, the incidence of severe grade III–IV aGVHD was not much higher than with a BM graft [64]. Nevertheless, it remains to be seen if outcomes with a PB graft are as good as with a BM graft in this setting. If the higher incidence of aGVHD has a negative impact on outcomes, an optimized PB graft will likely be needed.

3.2. Intensification of traditional GVHD prophylaxis

The Chinese investigators developed a different approach to control GVHD after HaploSCT. They used a myeloablative conditioning regimen, an intensified GVHD prophylaxis with ATG, cyclosporine, methotrexate, mycophenolate mofetil, and a donor graft composed of granulocyte–colony-stimulating factor (G-CSF)-primed BM and PB progenitor cells (GIAC protocol after G-CSF, intensified immunologic suppression, anti-thymocyte globulin, and combination of PB and BM grafts) [65]. Incidences of GVHD in 250 acute leukemia patients were higher than those seen with Post-Cy (46% grade II–IV aGVHD and 54% cGVHD), whereas almost all patients had successful engraftment. Di Bartolomeo et al. obtained similar results in Europe but reported a lower GVHD incidence using different myeloablative regimens and only a BM graft [66].

4. Haploidentical donor selection

Most patients requiring SCT have more than one haploidentical donor. The presence of recipient antibodies against donor-specific HLA, KIR mismatch predicting NK-cell alloreactivity, degree of HLA mismatch between donor and recipient, mismatch for noninherited maternal versus paternal alleles, donor age, and ABO-match may be important determinants of donor selection for HaploSCT.

Previous pregnancy or blood product transfusions may induce recipient anti-HLA antibodies against donor HLA antigens (DSA). The presence of DSAs is associated with increased risk of graft rejection [67]– [70]. Plasma exchange or rituximab may be used for recipients with DSA.

NK cells primarily attack hematopoietic cells sparing solid organs [19] and express inhibitory receptors, KIRs, that recognize epitopes shared by HLA class I alleles [16, 71]. In recipients lacking HLA class I alleles specific to the donor KIRs, donor NK cells may prevent GVHD and disease relapse by eliminating residual recipient antigen-presenting cells and leukemia cells [17, 72]. Accordingly, KIR mismatch between recipient and donor has been associated with improved HaploSCT outcomes with both TCD and T-cell replete grafts [17, 72]– [74]; however, this finding has been disputed by other researchers [75, 76].

Although a progressive increase in TRM with increasing genetic disparity has been historically reported, contemporary transplant strategies may negate this correlation by overcoming larger histoincompatibility barriers. In fact, Kasamon et al. and Wang et al. reported a similar incidence of acute GVHD and TRM after HaploSCT from full-haplotype mismatched donors compared with those from better-matched donors [77, 78].

The immune system is subject to senescence with advancing age. Accordingly, in the largest HaploSCT cohort published to date, Wang et al. reported a lower incidence of GVHD with younger donors compared with older ones [78]. Moreover, having a maternal donor was associated with a higher GVHD incidence and TRM than having a paternal donor. The latter is in contrast to the findings from a small registry study in which HaploSCT from maternal donors was found to be associated with lower TRM and longer OS compared with those from paternal donors [79]. Consequently, van Rood et al. demonstrated no significant differences in TRM, survival, or acute GVHD rates between HaploSCT from maternal and paternal donors [80]. The discrepancies between these studies are difficult to explain. However, both Wang et al. and van Rood et al. also found that HaploSCT from a sibling with a noninherited maternal antigen (NIMA) mismatch was associated with a lower GVHD incidence than that from a sibling with a noninherited paternal antigen (NIPA) mismatch supporting the hypothesis that the immunologic tolerance developed between the mother and the fetus during pregnancy [81, 82] may affect the transplant outcomes if the mismatched haplotype is of maternal origin. It is possible that although an immunologic tolerance is developed primarily in fetus against NIMA, immunity to minor histocompatibility antigen-encoded genes on the Y chromosome remains in the mother [83, 84]. Finally, older multiparous women may be the least preferred donors for male recipients [85].

Transplants involving a major ABO incompatibility require mononuclear cell separation to prevent a hemolytic reaction, which reduces the graft cell dose. If maximizing the infused stem cell dose is indeed important in HaploSCT, then younger, larger donors without a major ABO incompatibility with the recipient should be preferred.

With conflicting data, it is difficult to identify the optimal haploidentical donor. Until further evidence is available, we recommend the donor decision be based on age, NIMA mismatch, KIR mismatch, relation to the patient (mother the last choice), presence and level of anti-HLA antibodies, and ABO mismatch.

5. Outcome comparison with other transplant types

It was just over a decade ago when results from HaploSCT were significantly worse than those from matched and one-antigen mismatched unrelated donors [86]. Currently, the outcomes of HaploSCT are reported to be in par with those of transplants from HLA-matched donors. Among adults with intermediate- or high-risk acute myeloid leukemia in first complete remission, Wang et al. did not find any significant difference in survival, relapse rate, and TRM between transplants from HLA-identical siblings and haploidentical donors [87]. All transplants were performed with GIAC protocol except that ATG was not used in those from HLA-identical siblings. In another retrospective analysis, Raiola et al. reported a lower TRM with HaploSCT compared with cord blood and unrelated transplants and a longer survival compared with cord blood transplants [88]. In this cohort, Post-Cy and mostly ablative conditioning were used for HaploSCT. Kanda et al. reported worse survival and higher incidence of grade III–IV acute GVHD after HaploSCT compared with transplants from HLA-identical siblings [89]. However, in this study, HaploSCTs were performed with unmanipulated PB grafts and a GVHD prophylaxis including only alemtuzumab and mycophenolate mofetil without Post-Cy. Using the Post-Cy approach, Bashey et al. demonstrated similar outcomes between HaploSCT, transplants from matched related donors, and matched unrelated donors, with probabilities of disease-free survival of 60%, 53%, and 52%, respectively [90]. We have recently compared the outcomes of a uniform cohort of 227 patients with myeloid malignancies treated with the same conditioning regimen (fludarabine and melphalan) and found similar results. The 3-year disease-free survival for patients in complete remission after transplants from matched sibling, matched unrelated, and haploidentical donors were 51%, 45%, and 41%, respectively ($p = 0.4$) with similar immune reconstitution between the three groups [91].

6. Conclusion

Outcomes of HaploSCT have improved dramatically over the past several years, and its use has extended transplantation to virtually all patients in need. Although the optimal strategy to overcome the HLA–histoincompatibility barrier is debated, Post-Cy for GVHD prevention requires less resources and is associated with low TRM establishing itself as the new standard in HaploSCT. Novel methods for performing haploidentical transplants will have to be eventually compared with this approach. HaploSCT with Post-Cy has the potential to be the preferred transplant option for patients without HLA-matched donors worldwide, especially in developing countries where the cost of developing and maintaining unrelated donor registries or acquiring progenitor cells from the international registries might be prohibitive.

Author details

Stefan O. Ciurea[1*] and Ulas D. Bayraktar[1,2]

*Address all correspondence to: sciurea@mdanderson.org bv

1 Department of Stem Cell Transplantation and Cellular Therapy, The University of Texas MD Anderson Cancer Center, Houston, TX, USA

2 Division of Hematology, Memorial Sisli Hospital, Istanbul, Turkey

References

[1] Gragert L, Eapen M, Williams E, et al: HLA match likelihoods for hematopoietic stem-cell grafts in the U.S. registry. N Engl J Med 371:339–48, 2014

[2] Dehn J, Arora M, Spellman S, et al: Unrelated donor hematopoietic cell transplantation: factors associated with a better HLA match. Biol Blood Marrow Transplant 14:1334–40, 2008

[3] Passweg JR, Baldomero H, Bader P, et al: Hematopoietic SCT in Europe 2013: recent trends in the use of alternative donors showing more haploidentical donors but fewer cord blood transplants. Bone Marrow Transplant, 2015

[4] Luznik L, Jalla S, Engstrom LW, et al: Durable engraftment of major histocompatibility complex-incompatible cells after nonmyeloablative conditioning with fludarabine, low-dose total body irradiation, and posttransplantation cyclophosphamide. Blood 98:3456–3464, 2001

[5] Falk PM, Herzog P, Lubens R, et al: Bone marrow transplantation between a histocompatible parent and child for acute leukemia. Transplantation 25:88–90, 1978

[6] Dupont B, O'Reilly RJ, Pollack MS, et al: Use of HLA genotypically different donors in bone marrow transplantation. Transplant Proc 11:219–24, 1979

[7] Reisner Y, Kapoor N, Kirkpatrick D, et al: Transplantation for acute leukaemia with HLA-A and B nonidentical parental marrow cells fractionated with soybean agglutinin and sheep red blood cells. Lancet 2:327–31, 1981

[8] O'Reilly RJ, Kernan NA, Cunningham I, et al: Allogeneic transplants depleted of T cells by soybean lectin agglutination and E rosette depletion. Bone Marrow Transplant 3:3-6, 1988

[9] Aversa F, Tabilio A, Terenzi A, et al: Successful engraftment of T-cell-depleted haploidentical "three-loci" incompatible transplants in leukemia patients by addition of recombinant human granulocyte colony-stimulating factor-mobilized peripheral blood progenitor cells to bone marrow inoculum. Blood 84:3948–55, 1994

[10] Aversa F, Terenzi A, Tabilio A, et al: Full haplotype-mismatched hematopoietic stem-cell transplantation: A phase II study in patients with acute leukemia at high risk of relapse. J Clin Oncol 23:3447–54, 2005

[11] Hoffmann P, Ermann J, Edinger M, et al: Donor-type CD4(+)CD25(+) regulatory T cells suppress lethal acute graft-versus-host disease after allogeneic bone marrow transplantation. J Exp Med 196:389–99, 2002

[12] Nguyen VH, Shashidhar S, Chang DS, et al: The impact of regulatory T cells on T-cell immunity following hematopoietic cell transplantation. Blood 111:945–53, 2008

[13] Trenado A, Charlotte F, Fisson S, et al: Recipient-type specific CD4+CD25+ regulatory T cells favor immune reconstitution and control graft-versus-host disease while maintaining graft-versus-leukemia. J Clin Invest 112:1688–96, 2003

[14] Di Ianni M, Falzetti F, Carotti A, et al: Tregs prevent GVHD and promote immune reconstitution in HLA-haploidentical transplantation. Blood 117:3921–28, 2011

[15] Di Ianni M, Ruggeri L, Falzetti F, et al: HLA-Haploidentical Stem Cell Transplantation with Treg and Tcon Adoptive Immunotherapy promotes a Strong Graft-Versus-Leukemia Effect. Presented at the 2013 Annual ASH Meeting, 12/10/2013, 2013

[16] Vivier E, Tomasello E, Baratin M, et al: Functions of natural killer cells. Nat Immunol 9:503–10, 2008

[17] Ruggeri L, Capanni M, Casucci M, et al: Role of natural killer cell alloreactivity in HLA-mismatched hematopoietic stem cell transplantation. Blood 94:333–39, 1999

[18] Karre K, Ljunggren HG, Piontek G, et al: Selective rejection of H-2-deficient lymphoma variants suggests alternative immune defence strategy. Nature 319:675–78, 1986

[19] Asai O, Longo DL, Tian ZG, et al: Suppression of graft-versus-host disease and amplification of graft-versus-tumor effects by activated natural killer cells after allogeneic bone marrow transplantation. J Clin Invest 101:1835–42, 1998

[20] Koehl U, Esser R, Zimmermann S, et al: *Ex vivo* expansion of highly purified NK cells for immunotherapy after haploidentical stem cell transplantation in children. Klin Padiatr 217:345–50, 2005

[21] Passweg JR, Tichelli A, Meyer-Monard S, et al: Purified donor NK-lymphocyte infusion to consolidate engraftment after haploidentical stem cell transplantation. Leukemia 18:1835–38, 2004

[22] Nguyen S, Béziat V, Norol F, et al: Infusion of allogeneic natural killer cells in a patient with acute myeloid leukemia in relapse after haploidentical hematopoietic stem cell transplantation. Transfusion:no-no, 2011

[23] Yoon SR, Lee YS, Yang SH, et al: Generation of donor natural killer cells from CD34+ progenitor cells and subsequent infusion after HLA-mismatched allogeneic hematopoietic cell transplantation: a feasibility study. Bone Marrow Transplant 45:1038–46, 2009

[24] Choi I, Yoon SR, Park SY, et al: Donor-derived natural killer cells infused after human leukocyte antigen-haploidentical hematopoietic cell transplantation: a dose-escalation study. Biol Blood Marrow Transplant 20:696–704, 2014

[25] Ciceri F, Bonini C, Stanghellini MT, et al: Infusion of suicide-gene-engineered donor lymphocytes after family haploidentical haemopoietic stem-cell transplantation for leukaemia (the TK007 trial): a non-randomised phase I-II study. Lancet Oncol 10:489–500, 2009

[26] Di Stasi A, Tey SK, Dotti G, et al: Inducible apoptosis as a safety switch for adoptive cell therapy. N Engl J Med 365:1673–83, 2011

[27] Zhou X, Di Stasi A, Tey SK, et al: Long-term outcome and immune reconstitution after haploidentical stem cell transplant in recipients of allodepleted-T-cells expressing the inducible caspase-9 safety transgene. Blood, 2014

[28] Kochenderfer JN, Dudley ME, Carpenter RO, et al: Donor-derived CD19-targeted T cells cause regression of malignancy persisting after allogeneic hematopoietic stem cell transplantation. Blood 122:4129–39, 2013

[29] Mielke S, Solomon SR, Barrett AJ: Selective depletion strategies in allogeneic stem cell transplantation. Cytotherapy 7:109–15, 2005

[30] Amrolia PJ, Muccioli-Casadei G, Huls H, et al: Adoptive immunotherapy with allodepleted donor T-cells improves immune reconstitution after haploidentical stem cell transplantation. Blood 108:1797–808, 2006

[31] Chen BJ, Cui X, Liu C, et al: Prevention of graft-versus-host disease while preserving graft-versus-leukemia effect after selective depletion of host-reactive T cells by photodynamic cell purging process. Blood 99:3083–88, 2002

[32] Guimond M, Balassy A, Barrette M, et al: P-glycoprotein targeting: a unique strategy to selectively eliminate immunoreactive T cells. Blood 100:375–82, 2002

[33] Bastien JP, Krosl G, Therien C, et al: Photodepletion differentially affects CD4+ Tregs versus CD4+ effector T cells from patients with chronic graft-versus-host disease. Blood 116:4859–69, 2010

[34] Boussiotis VA, Gribben JG, Freeman GJ, et al: Blockade of the CD28 co-stimulatory pathway: a means to induce tolerance. Curr Opin Immunol 6:797–807, 1994

[35] Lin H, Bolling SF, Linsley PS, et al: Long-term acceptance of major histocompatibility complex mismatched cardiac allografts induced by CTLA4Ig plus donor-specific transfusion. J Exp Med 178:1801–06, 1993

[36] Guinan EC, Boussiotis VA, Neuberg D, et al: Transplantation of anergic histoincompatible bone marrow allografts. N Engl J Med 340:1704–14, 1999

[37] Davies JK, Gribben JG, Brennan LL, et al: Outcome of alloanergized haploidentical bone marrow transplantation after *ex vivo* costimulatory blockade: results of 2 phase 1 studies. Blood 112:2232–2241, 2008

[38] Sallusto F, Lenig D, Forster R, et al: Two subsets of memory T lymphocytes with distinct homing potentials and effector functions. Nature 401:708–12, 1999

[39] Lanzavecchia A, Sallusto F: Dynamics of T lymphocyte responses: intermediates, effectors, and memory cells. Science 290:92–97, 2000

[40] Ahmed R, Bevan MJ, Reiner SL, et al: The precursors of memory: models and controversies. Nat Rev Immunol 9:662–68, 2009

[41] Shlomchik WD: Graft-versus-host disease. Nat Rev Immunol 7:340–52, 2007

[42] Bleakley M, Otterud BE, Richardt JL, et al: Leukemia-associated minor histocompatibility antigen discovery using T-cell clones isolated by in vitro stimulation of naive CD8+ T cells. Blood 115:4923–33, 2010

[43] Anderson BE, McNiff J, Yan J, et al: Memory CD4+ T cells do not induce graft-versus-host disease. J Clin Invest 112:101–08, 2003

[44] Chen BJ, Cui X, Sempowski GD, et al: Transfer of allogeneic CD62L- memory T cells without graft-versus-host disease. Blood 103:1534-41, 2004

[45] Chen BJ, Deoliveira D, Cui X, et al: Inability of memory T cells to induce graft-versus-host disease is a result of an abortive alloresponse. Blood 109:3115–23, 2007

[46] Zheng H, Matte-Martone C, Li H, et al: Effector memory CD4+ T cells mediate graft-versus-leukemia without inducing graft-versus-host disease. Blood 111:2476–84, 2008

[47] Bender JG, Unverzagt K, Walker DE, et al: Phenotypic analysis and characterization of CD34+ cells from normal human bone marrow, cord blood, peripheral blood, and mobilized peripheral blood from patients undergoing autologous stem cell transplantation. Clin Immunol Immunopathol 70:10–18, 1994

[48] Bleakley M, Heimfeld S, Jones LA, et al: Engineering human peripheral blood stem cell grafts that are depleted of naive T cells and retain functional pathogen-specific memory T cells. Biol Blood Marrow Transplant 20:705–16, 2014

[49] Shook DR, Triplett BM, Eldridge PW, et al: Haploidentical stem cell transplantation augmented by CD45RA negative lymphocytes provides rapid engraftment and excellent tolerability. Pediatr Blood Cancer 62:666–73, 2015

[50] Vantourout P, Hayday A: Six-of-the-best: unique contributions of gammadelta T cells to immunology. Nat Rev Immunol 13:88–100, 2013

[51] Bonneville M, O'Brien RL, Born WK: Gammadelta T cell effector functions: a blend of innate programming and acquired plasticity. Nat Rev Immunol 10:467–78, 2010

[52] Godder KT, Henslee-Downey PJ, Mehta J, et al: Long term disease-free survival in acute leukemia patients recovering with increased gammadelta T cells after partially mismatched related donor bone marrow transplantation. Bone Marrow Transplant 39:751–57, 2007

[53] Schumm M, Lang P, Bethge W, et al: Depletion of T-cell receptor alpha/beta and CD19 positive cells from apheresis products with the CliniMACS device. Cytotherapy 15:1253–58, 2013

[54] Bertaina A, Pagliara D, Pende D, et al: Removal Of Alpha/Beta+ T Cells and Of CD19+ B Cells From The Graft Translates Into Rapid Engraftment, Absence Of Vis-

ceral Graft-Versus-Host Disease and Low Transplant-Related Mortality In Children With Acute Leukemia Given HLA-Haploidentical Hematopoietic Stem Cell Transplantation. Presented at the 2013 ASH Annual Meeting, 12/8/2013, 2013

[55] Berenbaum MC, Brown IN: Prolongation of homograft survival in mice with single doses of cyclophosphamide. Nature 200:84, 1963

[56] Jones RJ, Barber JP, Vala MS, et al: Assessment of aldehyde dehydrogenase in viable cells. Blood 85:2742–46, 1995

[57] Luznik L, Odonnell P, Symons H, et al: HLA-haploidentical bone marrow transplantation for hematologic malignancies using nonmyeloablative conditioning and high-dose, posttransplantation cyclophosphamide. Biol Blood Marrow Transplant 14:641–50, 2008

[58] Brunstein CG, Fuchs EJ, Carter SL, et al: Alternative donor transplantation after reduced intensity conditioning: results of parallel phase 2 trials using partially HLA-mismatched related bone marrow or unrelated double umbilical cord blood grafts. Blood 118:282–88, 2011

[59] Kasamon YL, Bolaños-Meade J, Gladstone D, et al: Outcomes Of Nonmyeloablative (NMA) Haploidentical Blood Or Marrow Transplantation (haploBMT) With High-Dose Posttransplantation Cyclophosphamide (PT/Cy) For Lymphoma, 2013

[60] Burroughs LM, O'Donnell PV, Sandmaier BM, et al: Comparison of outcomes of HLA-matched related, unrelated, or HLA-haploidentical related hematopoietic cell transplantation following nonmyeloablative conditioning for relapsed or refractory Hodgkin lymphoma. Biol Blood Marrow Transplant 14:1279–87, 2008

[61] Symons H, Chen A, Leffell M, et al: HLA-haploidentical bone marrow transplantation (BMT) for high-risk hematological malignancies using myeloablative conditioning and high-dose, posttransplantation cyclophosphamide. Blood 116:Abstract 2362, 2010

[62] Raiola AM, Dominietto A, Ghiso A, et al: Unmanipulated haploidentical bone marrow transplantation and posttransplantation cyclophosphamide for hematologic malignancies after myeloablative conditioning. Biol Blood Marrow Transplant 19:117–22, 2013

[63] Pingali SV, Denai M, Di Stasi A, et al: Haploidentical Transplantation For Patients With Advanced Hematologic Malignancies With Melphalan-Based Conditioning – Interim Results From a Phase II Clinical Trial, 2013

[64] Raj K, Pagliuca A, Bradstock K, et al: Peripheral blood hematopoietic stem cells for transplantation of hematological diseases from related, haploidentical donors after reduced-intensity conditioning. Biol Blood Marrow Transplant 20:890–95, 2014

[65] Huang X-J, Liu D-H, Liu K-Y, et al: Treatment of acute leukemia with unmanipulated HLA-mismatched/haploidentical blood and bone marrow transplantation. Biol Blood Marrow Transplant 15:257–65, 2009

[66] Di Bartolomeo P, Santarone S, De Angelis G, et al: Haploidentical, unmanipulated, G-CSF primed bone marrow transplantation for patients with high-risk hematological malignancies. Blood, 2012

[67] Ciurea SO, de Lima M, Cano P, et al: High risk of graft failure in patients with anti-HLA antibodies undergoing haploidentical stem-cell transplantation. Transplantation 88:1019–24, 2009

[68] Ciurea SO, Thall PF, Wang X, et al: Donor-specific anti-HLA Abs and graft failure in matched unrelated donor hematopoietic stem cell transplantation. Blood 118:5957–64, 2011

[69] Takanashi M, Atsuta Y, Fujiwara K, et al: The impact of anti-HLA antibodies on unrelated cord blood transplantations. Blood 116:2839–46, 2010

[70] Cutler C, Kim HT, Sun L, et al: Donor-specific anti-HLA antibodies predict outcome in double umbilical cord blood transplantation. Blood 118:6691–97, 2011

[71] Moretta L, Locatelli F, Pende D, et al: Killer Ig-like receptor-mediated control of natural killer cell alloreactivity in haploidentical hematopoietic stem cell transplantation. Blood 117:764–71, 2010

[72] Ruggeri L, Capanni M, Urbani E, et al: Effectiveness of donor natural killer cell alloreactivity in mismatched hematopoietic transplants. Science 295:2097–100, 2002

[73] Symons HJ, Leffell MS, Rossiter ND, et al: Improved survival with inhibitory killer immunoglobulin receptor (KIR) gene mismatches and KIR haplotype B donors after nonmyeloablative, HLA-haploidentical bone marrow transplantation. Biol Blood Marrow Transplant 16:533–42, 2010

[74] Mancusi A, Ruggeri L, Urbani E, et al: Haploidentical hematopoietic transplantation from KIR ligand-mismatched donors with activating KIRs reduces non-relapse mortality. Blood, 2015

[75] Huang Xj, Zhao Xy, Liu Dh, et al: Deleterious effects of KIR ligand incompatibility on clinical outcomes in haploidentical hematopoietic stem cell transplantation without in vitro T-cell depletion. Leukemia 21:848–851, 2007

[76] Gagne K, Brizard G, Gueglio B, et al: Relevance of KIR gene polymorphisms in bone marrow transplantation outcome. Hum Immunol 63:271–80, 2002

[77] Kasamon YL, Luznik L, Leffell MS, et al: Nonmyeloablative HLA-haploidentical bone marrow transplantation with high-dose posttransplantation cyclophosphamide: Effect of HLA disparity on outcome. Biol Blood Marrow Transplant 16:482–89, 2010

[78] Wang Y, Chang YJ, Xu LP, et al: Who is the best donor for a related HLA haplotype-mismatched transplant? Blood 124:843–50, 2014

[79] Tamaki S, Ichinohe T, Matsuo K, et al: Superior survival of blood and marrow stem cell recipients given maternal grafts over recipients given paternal grafts. Bone Marrow Transplant 28:375–80, 2001

[80] van Rood JJ, Loberiza FR, Jr., Zhang MJ, et al: Effect of tolerance to noninherited maternal antigens on the occurrence of graft-versus-host disease after bone marrow transplantation from a parent or an HLA-haploidentical sibling. Blood 99:1572–77, 2002

[81] Harris DT, Schumacher MJ, LoCascio J, et al: Immunoreactivity of umbilical cord blood and post-partum maternal peripheral blood with regard to HLA-haploidentical transplantation. Bone Marrow Transplant 14:63–68, 1994

[82] Claas FH, Gijbels Y, van der Velden-de Munck J, et al: Induction of B cell unresponsiveness to noninherited maternal HLA antigens during fetal life. Science 241:1815–17, 1988

[83] Randolph SS, Gooley TA, Warren EH, et al: Female donors contribute to a selective graft-versus-leukemia effect in male recipients of HLA-matched, related hematopoietic stem cell transplants. Blood 103:347–52, 2004

[84] Carlens S, Ringden O, Remberger M, et al: Risk factors for chronic graft-versus-host disease after bone marrow transplantation: a retrospective single centre analysis. Bone Marrow Transplant 22:755–61, 1998

[85] Ciurea SO, Champlin RE: Donor selection in T cell-replete haploidentical hematopoietic stem cell transplantation: knowns, unknowns, and controversies. Biol Blood Marrow Transplant 19:180–84, 2013

[86] Drobyski WR, Klein J, Flomenberg N, et al: Superior survival associated with transplantation of matched unrelated versus one-antigen-mismatched unrelated or highly human leukocyte antigen-disparate haploidentical family donor marrow grafts for the treatment of hematologic malignancies: establishing a treatment algorithm for recipients of alternative donor grafts. Blood 99:806–14, 2002

[87] Wang Y, Liu QF, Xu LP, et al: Haploidentical- versus identical-sibling transplant for AML in remission: a multi-centre, prospective study. Blood, 2015

[88] Raiola AM, Dominietto A, di Grazia C, et al: Unmanipulated haploidentical transplants compared with other alternative donors and matched sibling grafts. Biol Blood Marrow Transplant 20:1573–79, 2014

[89] Kanda J, Long GD, Gasparetto C, et al: Reduced-intensity allogeneic transplantation using alemtuzumab from HLA-matched related, unrelated, or haploidentical related donors for patients with hematologic malignancies. Biol Blood Marrow Transplant 20:257–63, 2014

[90] Bashey A, Zhang X, Sizemore CA, et al: T-cell-replete HLA-haploidentical hematopoietic transplantation for hematologic malignancies using post-transplantation cyclophosphamide results in outcomes equivalent to those of contemporaneous HLA-matched related and unrelated donor transplantation. J Clin Oncol 31:1310–16, 2013

[91] Di Stasi A, Milton DR, Poon LM, et al: Similar transplantation outcomes for acute myeloid leukemia and myelodysplastic syndrome patients with haploidentical versus 10/10 human leukocyte antigen-matched unrelated and related donors. Biol Blood Marrow Transplant 20:1975–81, 2014

Chapter 5

Cell Therapy in Huntington's Disease

Hyun Sook Kim and Jihwan Song

Additional information is available at the end of the chapter

http://dx.doi.org/10.5772/60618

Abstract

Huntington's disease (HD) is a rare neurodegenerative disease inherited in an autosomal dominant pattern. Expanded cytosine-adenine-guanine (CAG) repeats (polyQ) in the huntingtin gene cause the aggregates of abnormally expanded polyQ-containing huntingtin protein, and striatal medium spiny neurons are shown to be the most vulnerable. Affected patients develop cognitive, motor, and psychiatric symptoms typically in middle age, and several pharmacological drugs are currently used for symptomatic relief. Since the effect of current therapies is very limited and there is no way to modify disease progression, there is an unmet need for developing new therapies for HD. Toxin or genetic rodent models are widely used for drug development, and large animal models are also available. Previous studies transplanting cells originating from embryonic or fetal striatal tissues, neural stem cells, mesenchymal stem cells, and induced pluripotent stem cells (iPSCs) in HD animal models have shown the possibilities of clinical trials. Because clinical trials performed using human fetal striatal cells have shown variable outcomes, future directions of cell therapy in HD should consider the reconstitution of a functional dynamic information-processing circuit without ectopic connections. Another major challenge is to achieve controlled differentiation of embryonic stem cells or iPSCs into specific neuronal phenotypes.

Keywords: Huntington's disease, Animal models, Stem cells, Cell therapy, Transplantation

INTECH
open science | open minds

© 2015 The Author(s). Licensee InTech. This chapter is distributed under the terms of the Creative Commons Attribution License (http://creativecommons.org/licenses/by/3.0), which permits unrestricted use, distribution, and reproduction in any medium, provided the original work is properly cited.

1. Introduction

Huntington's disease (HD) is an inherited neurodegenerative disease caused by expanded CAG repeats in the huntingtin gene (*Htt*) on chromosome 4, which give rise to the formation of aggregates of mutant huntingtin proteins. Affected patients gradually develop cognitive decline, motor dysfunction (i.e., chorea or bradykinesia), and psychiatric disturbance that lead to progressive disability and death within approximately 15–20 years of disease onset. Since Huntington's disease is inherited in an autosomal dominant manner (i.e., 50 % of children are at risk) and the symptom onset occurs typically in middle age (median age, 35–40 years), affected patients and their family members suffer from significant economical and psychological burdens. Unfortunately, current therapies only target symptomatic reliefs and their effects are very limited, so the need for developing new therapies for HD is in high demand.

It is important to rescue these vulnerable medium spiny neurons (MSNs) by slowing the inexorable loss of striatal neurons and to delay the loss of striatal volume in affected patients, in view of HD pathology where striatal MSNs are shown to be most affected. More recently, stem cell strategy has been proposed to restore GABAergic striatal projection neurons into the putamen and the caudate, thereby reestablishing the degenerating striatopallidal circuit. According to previous research, transplanted cells originating from embryonic or fetal striatal tissues in HD animal models are connected with appropriate targets in the host brain and function both electrophysiologically and neurochemically to certain extents. Although clinical trials based on these preclinical studies have been performed using human fetal striatal cells, they have shown variable outcomes: some describe no benefit while others have indicated some clinical improvements with reduced motor dysfunction or slowed disease progression. However, because the use of human fetal tissues raise ethical issues and their genetic dissimilarity to the recipient is associated with the risk of immune rejection, other suitable non-fetal cell sources of syngeneic donor tissue would be advantageous. Recently, although induced pluripotent stem cells (iPSCs) generated from skin fibroblasts of HD patients have been demonstrated as an alternative autologous cell source, these HD patient-derived iPSCs carry genetic mutations, meaning that they have to be corrected to normal genes in order to be used for cell therapy. For clinical applications, it will be essential to use transgene-free reprogrammed iPSCs that are derived from patients under good manufacturing protocol (GMP) conditions. In addition to this iPSC approach, it may be worthwhile to consider using mesenchymal stem cells (MSCs), which are widely used in clinical trials, for the treatment of HD.

In this review, we describe the characteristics and limitations of current therapeutics and the need for developing novel ones. Then, animal models commonly used in HD research, various cell sources for transplantation, and the results and problems of preclinical and clinical studies so far are also discussed. Finally, we discuss the future directions of HD research and the clinical applications of stem cells. By doing so, we aim that the readers can acquire a thorough knowledge of HD and an understanding of the need for and the current view of cell therapy in HD.

2. Characteristics of Huntington's Disease

Huntington's disease (HD) is a progressively deteriorating neurodegenerative disorder caused by expanded polyglutamines (polyQs) with more than 35 CAG repeats in the huntingtin gene (*Htt*) on chromosome 4. Htt protein is approximately 350 kDa and is composed of ten HEAT (**h**untingtin, **e**longation factor 3, protein phosphatase 2**A**, and the yeast P13-kinase **T**OR1) repeats, which form α-helical structures participating in intracellular transport [1]. Htt protein is a ubiquitously expressed soluble protein [2]. Cytosolic Htt has been shown to bind microtubules and vesicles [3], whereas up to 5 % of Htt protein has been shown to locate within the nucleus [4]. Htt protein is known for its involvement in microtubule-mediated vesicle transport, endocytosis, secretory process, and regulation of gene transcription/RNA trafficking, in line with its location within the cell [5].

Interference with transcriptional regulation (e.g. CREB)
Decline of neurotransmitter receptors and brain-derived neurotrophic factor (BDNF) level
Interference with chaperone, proteasome (UPS), and autophagy activities
Disrupted axonal transport and synaptic transmission; cytoplasmic sequestration of transcription factors
Mitochondrial (complex I/II) dysfunction: calcium dysregulation/defective energy metabolism
Excitotoxicity via NMDA glutamate receptors
Activation of microglia and macrophages
Increased Kynurenine 3-monooxygenase (KMO) and 3-hydroxykynureine production: oxidative stress
Apoptosis/Necrosis

Table 1. Pathologic mechanisms of Huntington's disease

It is still under debate whether the effect of mutant Htt protein is through a "gain-of-toxic function" or a "loss of function," and various mechanisms have been suggested to elucidate the pathologic mechanisms of HD (Table 1). For example, mutant Htt that is translocated into the nucleus can interfere with gene transcription via either direct binding to DNA or interacting with several mediators including CBP (cAMP response element binding protein), NCoR (nuclear receptor corepressor), SP1 transcription factor, basal transcription factors, and REST (repressor element 1 silencing transcription factor) element [6, 7]. This reduced transcription can decrease the level of brain-derived neurotrophic factor (BDNF) in the brain of HD patients. Other loss-of-function mechanisms of Htt for disease pathogenesis have also been proposed [8, 9]. It is worth noting that Htt is necessary for early embryonic development and *Htt*-null mice demonstrate increased apoptosis [10]. A "gain-of-toxic-function" is another important mechanism of toxicity of mutant Htt. For instance, because a highly expanded *Htt* gene in HD leads to an Htt protein containing an abnormally expanded polyQ segment, toxic N-terminal fragments of abnormal ß-sheets are formed [11]. Other posttranslational factors also promote toxicity of mutant Htt, such as conformational change, aggregation propensity, cellular localization, and clearance rate. Mutant Htt binds to mitochondria and alters mitochondrial metabolism, which may result in energy defects, oxidative stress, and disturbed calcium homeostasis [12]. Moreover, protein clearance systems are shown to be impaired in HD patients and in models [13]. Irrespective of the mode of patho-mechanism, Htt aggregates within the cytoplasm and the nucleus, and selective neuronal cell loss and atrophy occur predominantly in the striatum and the neocortex [14]. Altogether, HD develops mainly through a "gain-of-toxic-function" mechanism from an abnormal conformation of mutant Htt [15, 16].

HD is an inheritable disease passed down in an autosomal dominant manner. The prevalence of HD is 4–10 per 100,000 in Western countries and approximately 0.4 per 100,000 in Asian countries [17]. Disease onset is typically in middle age (median age, 35–40 years), but it can also occur less commonly in juveniles and in old age depending on the CAG repeat number and/or modifying genes and environmental factors [18]. The studies on the relationship between CAG repeat number and disease manifestation demonstrated that there is an inverse correlation between age at initial symptom onset and the length of the expanded CAG repeat [19], and the correlation determines the age of onset with approximately 50–60 % variation [20, 21]. During HD, the disruption of the corticostriatal pathway, the main pathway affected, causes progressive cognitive decline, motor dysfunction, psychiatric disturbance, and ultimately death within approximately 15–20 years [22]. Although CNS degeneration attributes to these core symptoms, widespread pathology throughout the body may also contribute to other general symptoms such as weight loss, skeletal muscle wasting, metabolic and endocrine dysfunction, and dysfunction of cells of the hematopoietic lineage [13, 23].

Large aggregates of abnormally expanded polyQ-containing Htt protein, which form intranuclear inclusion bodies, are the pathological signatures of HD. Although the aggregates of mutant Htt are widespread throughout the brain and body, the striatum-selective damage encompassing the loss of striatal volume and up to 95 % loss of GABAergic medium spiny neurons (MSNs) is seen in the corticostriatal pathway [24]. This striatal selectivity might be explained by a possible involvement of a Ras homolog enriched in striatum (Rhes), a striatal-specific protein that binds to mutant Htt and increases the cytotoxicity of Htt through SUMOylation in HD pathogenesis [25]. Other than SUMOylation by Rhes, posttranslational modifications generally on the N-terminal 17 amino acids of Htt, including phosphorylation, ubiquitination, palmitoylation, and SUMOylation of huntingtin-associated (interacting) proteins, have important roles in modulating the toxicity of Htt as well as the selective neuronal loss [26]. Among the posttranslational modifications, phosphorylation has been shown as the major process for modulating Htt. Phosphorylation at serines 13 and 16 of Htt has been demonstrated to be protective against motor and psychiatric dysfunction and neuropathology [27]. Palmitoylation of Htt at cystein 214 has been shown to enhance its membrane association, whereas an expanded Htt shows less palmitoylation and an associated increase in neuronal toxicity [28]. Small ubiquitin-like modifiers (SUMO) are proteins that covalently attach to and detach from the target protein to modify their function, a process described as SUMOylation. Several SUMO proteins are known to interact with Htt-related proteins (huntingtin-associated protein 1 (HAP1) and transcription elongation regulator 1) [13], and SUMOylation of these proteins by SUMO may be related to the pathogenesis of HD. Lastly, a disruption of BDNF support or an increased susceptibility to glutamatergic excitotoxicity of the neuronal circuit between the cerebral cortex and the striatum can lead to HD pathogenesis [29].

In HD pathogenesis, selective striatal neuronal loss comes not only from cell-autonomous toxicity but also cell-cell interactions. Cell-cell interactions occur between both interneuronal connections and between glial cells and neurons. In interneuronal connections, mutant Htt causes increased stimulation of extrasynaptic glutamate receptors and a decreased reuptake of glutamate by glia, leading to enhanced excitotoxicity and metabolic toxicity. Neuron and glia coculture experiments showed that expressing mutant Htt in glia triggered neuronal death only in cells expressing mutant Htt, providing evidence for the role of mutant Htt in neuronal excitotoxicity [30]. Further, HD patient-derived astrocytes alone were shown to replicate HD pathology when Htt with expanded CAG repeats was expressed [31]. Similarly, in astrocytes, transgenic expression of mutant Htt alone led to HD-like symptoms or worsened disease progression when crossed into existing HD models or even into normal mice [32]. In HD patients, microglia are activated in prodromal stages, and then symptoms manifest in HD. The level of microglial activation has been shown to correlate with disease severity and striatal loss *in vivo* [33, 34].

3. Current therapeutics in clinical practice and its limitation: Need for developing novel therapeutics

Several pharmacological drugs are currently used for symptomatic relief of HD symptoms such as hyperkinetic involuntary movements and mood disorders, and there is no way to modify disease progression. For hyperkinetic involuntary movements including chorea, myoclonus, and dystonia, dopamine-depleting agents (tetrabenazine), antipsychotics (haloperidol, pimozide, clozapine, olanzapine, ziprasidone, aripiprazole, risperidone, and quetiapine), benzodiazepines (clonazepam), anticonvulsants (sodium valproate and levetiracetam), and botulinum toxin can be prescribed. Because tetrabenazine is not only a dopamine-depleting agent but may also decrease brain serotonin and norepinephrine concentrations, it potentially causes or aggravates depression [35]. Various neuroleptics (olanzapine, quetiapine, risperidone, sulpiride, haloperidol, and clozapine) that are also used for psychiatric symptoms should be administered with caution because they may induce tardive dyskinesia and other adverse effects. Clozapine is well known for being costly and inducing irreversible agranulocytosis. For mood disorders, antidepressants (citalopram, fluoxetine, paroxetine, sertraline, mirtazapine, and venlafaxine) and anxiolytics (benzodiazepines, propranolol, and clonidine) are used [13]. A range of psychiatric drugs can alleviate some of the more overt disturbances of mood and hyperactivity in HD; however, these drugs have limited effects and are associated with side effects [36]. There is limited evidence in the literature for the use of acetylcholinesterase inhibitor (rivastigmine and donepezil) for cognitive dysfunction in HD [36, 37].

Behavioral and social interventions are often as effective as drug therapy for behavioral difficulties. For instance, weight loss frequently leads to general weakening [38] and a higher body mass index has been associated with a slower rate of disease progression [39], meaning that adequate nutrition may be another important aspect of therapy. Other than nutrition, environmental enrichment and physical, speech, and occupational therapy are also usually recommended to delay the onset of HD, some of which have been validated in mouse models [40, 41].

HD is one of few neurodegenerative diseases for which the diagnosis can be made long before the onset of clinical symptoms by predictive genetic testing. This offers an opportunity to intervene in the earliest stages of neurodegeneration and thereby slow down or arrest disease progression. Many drugs targeting various mechanisms underlying HD pathogenesis have been tested. However, antiexcitotoxic drugs, such as riluzole (an antiglutamatergic drug) [42], vitamin E [43], idebenone and remacemide (an NMDA ion channel blocker) [44], minocycline (a caspase-3 inhibitor and anti-inflammatory agent) [45], and creatine (energy metabolites, CREST-E trial) [46] have failed in clinical trials so far. More favorable results from animal studies using memantine (an NMDA receptor antagonist), coenzyme Q10 (a mitochondrial cofactor and an antioxidant), and ethyl eicosapentaenoate (an antioxidant) have been highlighted, but none of these compounds was successful. Currently, reducing the expression of mutant Htt protein with RNA interference or antisense oligonucleotides is the most promising candidate and other approaches of various mechanisms are being designed and under investigation [47].

4. Animal models of Huntington's Disease

4.1. Rodent models in Huntington's Disease

Rodent models have played an important role in providing experimentally accessible systems to study various clinicopathological findings and pathogenesis of HD and to test potential therapeutics of their efficacy [48]. From the late 1970s, several investigations started to generate animal models of HD. Glutamate-related excitotoxin kainic acid (KA) and quinolinic acid (QA) were used to induce the degeneration of striatal GABAergic projection neurons while preserving striatal afferents, thereby producing a model that resembles the neuropathologic condition in HD [49, 50]. Because of marked epileptogenic side effects in KA model, QA model is favored and currently used. Another toxic model of HD is made by producing defective energy metabolism through toxins such as sodium azide, malonate, and 3-nitropropionic acid (3-NP). For example, 3-NP-induced inhibition of mitochondrial complex II effectively produces striatal lesions that were similar to the cell loss in HD [51, 52]. To mimic the chronic progressive nature of human HD, chronic administration of metabolic toxin has been tried in experimental primates; however, higher inter-animal variability and nonspecific striatal damage in primates (NO replacement; just delete please) limit this approach [53]. Altogether, despite their usefulness as mimics of striatal pathology and behavioral manifestations of HD, toxin models were limited because it was not possible to study disease progression or to replicate the widespread neuropathology observed in the human condition [54].

After the discovery of the HD gene in 1993, genetic models of HD have been generated, and over 20 different rodent models have been generated to date [55]. Genetic models provide accurate and experimentally accessible systems in which to study the molecular pathogenesis. Moreover, they provide an opportunity to test the effect/efficacy of candidate therapeutics and explore their potential for clinical applications. Because the degree of overexpression of mutant Htt protein plays a significant role in the phenotype in mice, genetic mouse lines have been generated with varying degrees of phenotype by incorporating variations of mutant huntingtin gene into the mouse genome. The mouse models fall into three categories (Table 2): (i) mice that express truncated N-terminal fragment (exon-1 or exons 1 and 2) of the human *Htt* gene containing polyglutamine mutations [56-60]; (ii) mice that express the full-length human HD gene [61-65]; and (iii) mice with pathogenic human CAG repeats inserted into the existing CAG expansion in murine *Htt* (knock-in models) [66-70]. Although all of these models develop the typical findings of human HD, the degree and the progression to which the behavioral features and pathological findings manifest differ, as well as the developmental speeds. Among them, the R6/2 mice showed the most prominent motor, behavioral, and cognitive phenotypes as well as marked weight loss and death by 13~15 weeks of age. In this strain, various mechanisms of HD pathogenesis are demonstrated, including the intraneuronal nuclear inclusions; impaired mitochondrial function [71]; abnormalities of glutamatergic, dopaminergic, and cholinergic receptors in the striatum [72]; and abnormalities in synaptic plasticity in the hippocampus [73, 74]. The full-length *Htt* gene (transgene) is incorporated into the mouse genome via bacterial artificial chromosome (BAC) or yeast artificial chromosome (YAC). Meanwhile, the phenotypes in the full-length *Htt* mutation model develop gradually over several months and may survive as long as the wild-type animal [75]. This model is especially valuable for studies on the proteolytic mechanism of full-length *Htt*, its clearance, and the evaluation of the presymptomatic stages of HD. Among the full-length HD genetic models, the YAC mouse model with 128 CAG repeats (YAC128) develops motor abnormalities, composing of an initial hyperactivity and followed by difficulty in motor control from six to 12 months and then hypokinesia [76]. Further, other motor dysfunctions, including circling behavior, hind limb clasping, and gait abnormalities, may be seen as early as three months [77]. Although the full-length models are genetically more accurate, the fragment *Htt* models have been used more frequently for their aggressive phenotype, rapidly progressive disease course, well-defined behavioral and pathological findings, and early death. Moreover, due to the prolonged disease progression in the full-length *Htt* models, it is hardly possible to use progressive morbidity and survival as endpoints [53].

While toxin models play a role in understanding mechanisms of excitotoxicity and mitochondrial dysfunction in HD, they cannot replicate the progressive neurodegenerative course characterized by the misfolding of the mutant Htt protein in HD. Alternatively, genetic animal models provide a good platform to explore the progressive manifestation of neuropathology and cognitive, behavioral, and motor dysfunction [48]. It also provides the platform to test potential therapeutics for future translational research. However, standardization of sample size, inclusion/exclusion criteria of mice, and the onset and duration of treatment, as well as the outcome measurement of preclinical trials, are critical to compare the effectiveness of candidate therapeutics [78].

Model	CAG-N	Behavioral changes	Neuropathology	Survival
Transgenic models: Truncated N-terminal fragment				
R6/1, Mouse [56, 57]	116	Decreased anxiety Significant weight loss Abnormal motor performance	Reduced brain volume Loss of striatal neurons Htt aggregates Reduced dopamine levels	12+ months
R6/2, Mouse [58]	144–150	Dystonia with limb clasping Significant weight loss Reduced motor performance Seizures, Diabetes	Gross brain atrophy Progressive neuronal atrophy with neuronal loss Htt aggregates Astrogliosis Reduced dopamine levels	12–18 weeks
N171-82Q, Mouse [59]	82	Weight loss Abnormal motor performance Limb clasping Visuospatial memory loss	Gross brain atrophy Atrophy and loss of striatal neuron Htt aggregates	130–180 days
HD51, Rat [60]	51	Significant weight loss Deficient motor performance Reduced anxiety Cognitive deficits Head dyskinesias	Enlarged ventricles Striatal neuronal loss Htt inclusion	NA
Transgenic mouse models: Full-length human *Htt*				
YAC 128 [62, 76]	128	Hyperactivity (initial); hypokinesia (later) Abnormal circling behavior Hindlimb clasping Deficient motor performance Gait abnormalities	Decreased striatal and cortical volume Reduced striatal neuron area and number Progressive Htt aggregates	Slightly decreased
BACHD [64, 78]	97 (Mixed CAA/CAG repeat)	Significant reduction in motor function (2 months) Behavioral worsening (12 months)	Marked gross brain atrophy and brain weight loss Significant cortical and striatal volume loss (12 months) Degenerating neurons in striatum (12 months) Htt inclusions in entire cortex; a few small inclusions in striatum	Normal lifespan
Hu97/18 [65]	97	Motor learning deficit (2 months) Deficits in both spatial learning and object recognition (9 months) Increased stereotypy or repetitive movement Anxious and depressive-like behaviors	Forebrain atrophy Striatal volume loss Cortical shrinkage and white matter loss (12 months)	NA

Model	CAG-N	Behavioral changes	Neuropathology	Survival
Knock-in mouse models: Full-length *Htt*				
HdhQ111 [66, 68]	111	Gait impairment (13 months)	Diffuse Htt activity (6 weeks) with nuclear inclusions (12 months) Astrogliosis (24 months)	Normal life span
CAG140 [69]	140	Body weight loss Hyperactivity (initial); hypoactivity (later) Gait abnormalities (12 months)	Nuclear and neuropil aggregates (8 months) Diffuse Htt (2 months) Neuronal loss	Normal life span
CAG150 [70]	150	Body weight loss (70 weeks) Motor performance deficits on rotarod, gait, and beam balance (70–100 weeks)	Striatal Htt aggregates (28 weeks) Nuclear inclusions (37 weeks) Reactive astrogliosis (56 weeks) Loss of striatal neuron perikarya and volume (100 weeks in homozygous mice)	Normal life span

Htt=huntingtin protein; NA=not available

Table 2. Rodent models of Huntington's disease

4.2. Large animal models in Huntington's Disease

Nonhuman primates are genetically more similar and have a more similar physiology to humans than rodents, making them invaluable for modeling human disorders and for developing therapeutic strategies. Nevertheless, only a limited number of works has been reported in HD. The use of nonhuman primates is focused on the study of HD-like behavioral manifestations, especially for chorea, and the development of potential therapeutics for HD. In the case of nonhuman primate study, the toxin models using QA or 3-NP were most commonly used [52, 79]. After the development of transgenic nonhuman primate models of HD in 2008, the potentials of using large animals in HD research have been spotlighted. There are three types of large animal models: a rhesus monkey (Macaca mulatta) [80], a miniature pig model [81], and a sheep model (Ovis aries) [82]. The rhesus macaque and pig models were generated using fragments of human mutant *Htt* and the sheep model using the full-length human coding sequence of *Htt*. By injecting oocytes with lentiviruses expressing exon1 of *Htt* carrying 84 CAG repeats, three transgenic monkeys were made, which survived for more than two years and showed clinical features of HD, including dystonia and chorea [80]. Transcriptomic dysregulation was recently reported from peripheral blood samples, which is under further clinical investigations [83]. The transgenic HD sheep model demonstrated a decreased expression of DARPP-32 (dopamine- and cAMP-regulated phosphoprotein of 32 kDa) at seven months of age and a characteristic inclusion pathology in the brain at 18 months [84].On the contrary, there were no aggregates in the brain even up to 16 months of age and no development of motor abnormalities in transgenic minipigs [85]. Because transgenic large animal models have many advantages to understand the biology of HD and the development of potential therapies, practical and ethical issues as well as preclinical works should be also considered.

5. Source of cell therapy for striatal transplantation in HD

A characteristic pathological feature in Huntington's disease is a selective loss of medium spiny projection neurons (MSN) in the striatum. Thus, cell therapy in HD is aimed at replacing the MSN and making them functionally active by connecting them to the host neuronal network [86]. To accomplish this, transplanted cells should survive, differentiate in vivo into the proper cell type (i.e., MSN), be functionally active, and connect with appropriate target neurons, thereby reestablishing the degenerating striatopallidal circuit [87, 88]. A number of potential stem/progenitor cells have been studied that include embryonic stem cells; multipotent progenitor cells from the embryo or fetus, which are already partially committed to a neural lineage; cells from the umbilical blood; autologous or allogenic adult stem cells from various tissues; and finally induced pluripotent stem cells [53]. The majority of cells used so far are fetal neural stem/progenitor cells, but adult stem cells or iPSCs can be alternative cell sources to fetal or embryo-derived stem cells. Before starting clinical trials, each cell type should be shown to have efficacy and long-term safety in properly designed animal models of HD.

5.1. Fetal neural stem cells and fetal neural progenitor cells

Fetal neural stem cells (NSCs) are isolated from the fetal brain at various gestational periods and from multiple brain regions [89]. After the collection of primitive cells and their short-term expansion in vitro, these cells are transplanted into the lesioned brain of an adult rodent. Majority of previous studies have used fetal neural progenitors (FNPs), which are expanded as neurospheres prior to transplantation (Table 3), and the transplanted FNPs showed evidence of differentiation into striatal-like cells [90]. Such neural differentiation depends on the characteristics of fetal neural stem/progenitor cells to respond to signals in the developing CNS [91]. For example, stem cells derived from the human fetal cortex (12 weeks post-conception) were pretreated in culture media with ciliary neurotrophic factor (CNTF) and were transplanted into QA-lesioned rats. Motor recovery and sustained striatal volume were marked and the transplanted human stem cells differentiated into neurons and astrocytes with substantial connections to endogenous cells [92]. Systemic injection of human NSC also improved neuropathologic and behavioral abnormalities [93]. Because HD showed a selective loss of MSNs in the striatum, induction of a GABAergic phenotype in immortalized striatal NSCs was tried. These GABAergic phenotype-induced cells maintained neurite processes connecting to endogenous neurons [94]. Mouse NSCs modified to secrete human nerve growth factor (NGF) were transplanted into striatal QA-lesioned rats, and it was shown that the size of lesion, the number of surviving striatal neurons, and the length of neurites were significantly improved than sham-operated rats [95]. However, despite these successes based on fetal stem/progenitors, the use of in vitro expanded fetal neural stem cells is limited because they lose the capacity to differentiate into various neural cell types and tend to go into senescence after several passages of culture [96]. Moreover, there is an intrinsic ethical issue associated with the use of aborted fetal tissue. Recently, a cloned human striatal neural stem cell line (STROC05) was transplanted in the R6/2 mouse model of HD, but the mouse model showed a disappointing suboptimal clinicopathological improvement [97] (Table 3).

Animal model	Cells	Histology	Functional outcome	Ref
Rat, QA [U]	Human stem cell-derived from fetal cortex (pretreated with CNTF)	Reduced striatal atrophy Survive and differentiate into neuron and glia Connectivity with endogenous neural cells	Improved motor performance (cylinder test)	[92]
Rat, QA [U]	Human NSC-derived from fetal human brain (systemic injection)	Reduced striatal atrophy NSC migration in and around the damaged striatum Migrated NSCs differentiated into neurons and glias	Reduced ApoM-induced rotation	[93]
Mice, QA [U] R6/2	Mouse NSC - neurosphere and dissociated cell suspension	Increased survival of graft when transplant at 2 days after lesioning	No change on BDNF expression	[94]
Rat, QA	Noggin-primed human NPC derived from human ESC	Extensive migration and large-scale differentiation Increased the extent of neuronal differentiation	NA	[102]
R6/2	undifferentiated or predifferentiated DARPP-32 cells [B] derived from human striatal neural stem cell line (STROC05)	Poor survival and neuronal differentiation both in the undifferentiated and differentiated conditions A few cells expressed the neuronal marker beta-III-tubulin.	NA	[97]
Rat, Lesions of the dorsal striatum [U]	Homotopic neural transplants (GE or cortex from E15 rat embryos of same strain)	Patches of positive DARPP-32 and tyrosine hydroxylase Significantly higher extent of DARPP-32 patches	Alleviated lateralised response deficits Prevented development of lateral disparity Restored speed of responding back to pre-lesion levels	[131]
Rat, KA	Fetal rat striatal primordia	Differentiation to spiny neuron	Restored synaptic potential	[135]
Rat, QA	Embryonic striatal grafts (Lateral GE)	Reversed lesion-induced increase in the cytochrome oxidase activity of the Gp	Reduction of ApoM-induced rotational asymmetry	[136]
Rat, Ch	Embryonic striatum	Innervated by host-derived dopamine axons	Restored response of host neurons	[137]

Animal model	Cells	Histology	Functional outcome	Ref
Rat, KA	Fetal striatal graft	NA	Restored striatal GABA overflow	[138]
Rat, IA [U]	Embryonic striatal graft	NA	Stimulation of GABA release	[139]
R6/2	Striatal grafts	Survival and differentiation of grafts	Improved general locomotor behavior No significant functional improvement	[132]
R6/1	Wild-type donor cortex	NA	No improvement	[133]
Rat, 3-NP	Human NSC (transplantation before 3-NP administration)	Reduced damage to striatal neurons Increased BDNF expression	Improved motor performance	[140]

Apo-M=apomorphin; BDNF=brain-derived neurotrophic factor; BM-MSCs=bone-marrow-derived mesenchymal stem cells; Ch= Cholecystokinin-8-sulphate; CNTF=ciliary neurotrophic factor; DARP-32=dopamine- and cAMP-regulated phosphoprotein of 32 kDa; ESC=embryonic stem cell; GABA=gamma-aminobutyric acid; GE=ganglionic eminence; Gp=globus pallidus; IA=ibotenic acid; KA=kainic acid; 3NP=3-nitropropionic acid; MSN=medium spiny neuron; NA=not available; NPC=neural precursor cell; NSC=neural stem cell; QA=Quinolinic acid; [B]=bilateral; [U]=unilateral

Table 3. Neural stem cell/progenitor cell-based treatment of HD in preclinical rodent models

5.2. Embryonic stem cell-derived neural progenitor cells

Embryonic stem cells (ESCs) are isolated and expanded from the inner cell mass of the blastocyst-stage embryo. Because ESCs are very primitive and pluripotent, they can be expanded in vitro indefinitely while retaining relatively stable cell characteristics [98]. Even after expansion, they also retain substantial neurogenic potential [99]. Their ability to expand and sustain neurogenic potential provides many advantages to be used in cell therapy; however, there have been difficulties in inducing differentiation of ES cells into striatal cells, which is especially important in HD. A report showed that mouse ES cells treated with retinoic acid could differentiate into neuronal cells that could integrate and survive in the QA-lesioned rat model of HD [100]. Another concern with the use of these cells is its potential for tumorigenicity, because even tiny numbers of undifferentiated cells at the time of transplantation may develop into tumors later on. To overcome these limitations of ES cells, research has shifted to using neural stem/precursor cells (NPCs). NPCs are a heterogeneous population of mitotically active, self-renewing, and multipotent cells, which can be isolated from the embryo [101]. Likewise, to use human neural precursors (hNP) derived from embryonic stem cells (hESCs) in HD cell therapy, hNP is required to differentiate into neuronal cells, especially MSN, in vivo. In vitro noggin priming can be an effective tool of hNP transplantation in HD treatment. Noggin-primed hNP showed survival, extensive migration, and differentiation into predominantly neuronal cells after transplantation in the QA-lesioned stratum of rats [102]. Recently, a protocol has been developed to obtain a high percentage of functional GABAergic neurons from mouse embryonic stem cells (mESCs) [103]. By combining an in vitro culture and an in vivo differentiation protocol, striatal progenitors derived from hESCs were shown to mature into DARPP-32-positive neurons in QA-lesioned rats [104]. Using neural precursor cells with elimination of undifferentiated cells, ESC-based regenerative approaches may be successful in treating HD patients. In addition to cell replacement potential, there is evidence that transplantation of NPCs may modulate inflammatory reactions through a "bystander" mechanism [105].

5.3. Non-neural stem cells

To avoid ethical problems, non-neural stem cells, such as adult stem cells or umbilical cord-derived cells, have been used as alternative cell sources for HD treatment. Because adult stem cells are relatively easy to harvest and autologous grafting is possible, there have been many studies using mesenchymal stem cells (MSCs) of the bone marrow [106-108] or adipose tissue [109, 110] in HD mouse models (Table 4). Although there is still debate about the possibility of cell fusion events rather than true transdifferentiation, these non-neural stem cells are known to have the capacity to differentiate into neurons after injection into the adult rodent host [111]. Intrastriatal MSC transplants reduced motor and pathological deficits in a 3-NP-lesioned rat and QA-lesioned mouse/rat HD models [106, 109, 112-115]. Genetically engineered stem cells that produce trophic factors could also be a cell source for cell therapy [116-118]. When genetically engineered bone marrow-derived MSCs overexpressing neurotrophic factors (NTFs), called NTF(+) cells, were transplanted into rat brains after QA injection, NTF(+) cells survived, maintained their NTF-secreting phenotype, and exhibited improved behavior and reduced striatal atrophy associated with QA lesions [119]. MSCs engineered to produce BDNF also improved behavioral performance and reduced striatal atrophy when transplanted in YAC128 mouse model of HD [120]. In addition, MSC-based studies take advantages of the property of MSCs to modulate the brain environment toward neuroprotection [116]. Injection of cell-free extracts of adipose-derived stem cells also demonstrates behavioral and pathological improvements of R6/2 models in terms of HD pathology [121].

Animal model	Cells	Histology	Lesion volume	Functional outcome	Ref
Rat, QA	Rat BM-MSCs	Reduced striatal atrophy	Improved striatal volume	NA	[106]
Rat, QA	Human adipose MSCs	Reduced striatal atrophy and apoptosis	Decreased lesion volume	Reduced Apo-M induced rotations	[109]
Mouse, QA	Human BM -MSCs	Increased cell proliferation in striatum Reduced apoptosis	Decreased lesion volume	Improved rotarod performance Extended survival	[113]
R6/2 mouse	Human BM -MSCs	Improved cell differentiation	NA	Improved survival	[113]

Animal model	Cells	Histology	Lesion volume	Functional outcome	Ref
Rat, QA	Human BM -MSCs	Reduced striatal atrophy Increased level of NTFs	Decreased lesion volume	Reduced motor dysfunction	[114]
R6/2 mouse	Human adipose MSCs	Reduced Htt aggregates Attenuated loss of striatal neurons	Improved striatal volume	Improved rotarod Reduced clasping Improved survival	[109]
Rat, 3NP	Rat BM-MSCs	Increased striatal labeling in BDNF, collagen type-I and fibronectin	Prevented 3NP-mediated ventricle enlargement	Improved rotarod Improved paw placement	[112]
Rat, QA	Rat BM-MSCs	Improved MSC migration to lesion	Decreased lesion volume	Regenerated striatal network Reduced Apo-M induced rotations	[107]
Rat, QA	Human BM-MSCs engineered to secret NTF	Survived after 6 weeks Sustained NTF secretion	Decreased lesion volume	Reduced Apo-M induced rotations	[119]
YAC 128 mouse	Human adipose MSCs	Reduced striatal atrophy	Improved striatal volume	Improved rotarod performance/ motor function	[110]
Rat, QA	Rat BM-MSCs	NA	NA	Reduced Apo-M induced rotations Improved beam walk and hang wire time	[108]
YAC 128 mouse	BM-MSCs engineered to produce BDNF	Reduced striatal atrophy	Improved striatal volume	Improved rotarod performance Reduced hindlimb clasping	[120]
Mouse, QA	hESC-derived GABA neurons and their progenitors	Repopulated GABAergic cells, connecting with endogenous cells	NA	Improved rotarod, openfield, and tradscan performance	[142]

Apo-M=apomorphin; BDNF=brain-derived neurotrophic factor; BM-MSCs=bone-marrow-derived mesenchymal stem cells; hES=human embryonic stem cells; NA=not available; NTF=neurotrophic factor; QA=quinolinic acid; 3NP=3-nitropropionic acid

Table 4. Mesenchymal stem cell-based treatment of HD in preclinical rodent models

5.4. Induced Pluripotent Stem Cells (iPSCs)

Induced pluripotent stem cells (iPSCs) generated from somatic cells of patients can be used to model different human diseases, thereby enabling disease investigation and drug development [122]. Since the generation of iPSCs does not involve the destruction of human embryos, they can avoid the ethical issues related to the use of human ESCs [123]. In direct opposition to embryonic/fetal tissue-derived cells, patient-derived iPSCs can avoid immune rejection as well as ethical problems, and they can also serve as sources of transplantable cells in novel cell therapies [124]. Recently, non-integrating episomal vectors were introduced into the cell by electroporation, making transgene-free and virus-free iPSCs in a feeder-free environment [125]. There are controversial results regarding formation of mutant Htt proteins in HD patient-derived iPSCs (HD-iPSCs) [126, 127]. At least, HD-iPSCs can be differentiated into GABAergic striatal neurons and make significant behavioral recovery without aggregation formation at 12 weeks after transplantation [127]. Nevertheless, as HD is a genetic disorder, correction of HD mutations in HD-iPSCs is suggested to be an essential step before grafting to HD patients [122]. With the advent of new gene editing technologies using zinc-finger nuclease (ZFN), transcription activator-like effector nuclease (TALEN), and clustered regularly interspaced short palindromic repeats (CRISPR)/Cas9, iPSCs that are derived from patients with gene mutations may still be a plausible cell source for transplantation [128, 129]. A nucleofection-based protocol adapted to single-cell dissociated feeder-free culture was established, and a reversal of HD phenotype and striatal differentiation of gene-corrected HD-iPSCs *in vitro* and *in vivo* are demonstrated [130]. If transgene-free reprogrammed iPSCs derived from patients could be made under good manufacturing practice (GMP), they may provide a suitable source for autologous transplantation in the future.

6. Results and problems of pre-clinical and clinical studies

6.1. Pre-clinical cell transplantation in Huntington's Disease models

In view of HD pathology, it is important to rescue the vulnerable MSNs by slowing the inexorable loss of striatal neurons. To rescue the damaged striatum, embryonic striatal tissues were transplanted, and their survival, expression of a wide range of striatal markers, and the recovery of motor and cognitive functions were demonstrated by many tasks of evaluation both in HD toxin and genetic models (Tables 3–4) [131-133]. Although the extent of functional improvement was relatively modest in transgenic animal models, it was shown that the transplanted cells can connect with appropriate targets in the host brain [134] and function both electrophysiologically [135, 136] and neurochemically [137, 138], indicating that striatal grafts can yield a functional repair of striatal cell loss in HD [139]. The timing of cell transplantation should be also considered, as the pretreatment with human NSCs was shown to rescue the motor impairment and neuronal damage caused by systemic 3-NP administration [140]. When NSCs are transplanted after excitotoxin injection, the extent of neuronal recovery could be different according to the time interval between excitotoxin injection and NSC transplantation [94]. Another consideration on the cell transplantation is that transplanted cells should exhibit an MSN morphology, express MSN markers such as DARPP-32, and function *in vivo* to recover the lost functions [141]. Recently, hESC-derived DARPP32-expressing forebrain GABA neurons and their progenitors were implanted into QA-lesioned mice and they were connected with endogenous dopaminergic and glutamatergic neurons [142]. In the past two decades, clinical trials have been performed using human fetal striatal cells, and some reports have indicated clinical improvement with reduced motor dysfunction or slowed disease progression in some patients (Table 5). However, because human fetal tissue transplants inevitably raise ethical issues and risk of immune rejection, induced pluripotent stem cells (iPSCs) generated from skin fibroblasts of HD patients [122], can serve as an autologous cell source. HD-iPSCs could be differentiated into neural precursor cells and implantation of those cells in the striatum of QA-lesioned rats; behaviors improved significantly; and Htt aggregation was not formed at 12 weeks after transplantation [127]. In addition, various adult stem cells also demonstrated behavioral and pathological improvement after transplantations into toxin or transgenic rodent models (Table 4).

Donor tissue	N	Implant site	Safety	Efficacy	Ref
VM or WGE	4	NA [B]	No pathological or immunological response	Not possible to determine	[143]
E12-13 WGE	2	CN cavity	No surgical incidents or subsequent SEs	Slow progression	[144]
E8-9 LLGE	7	pcPu [B]	1 death from cardiac arrest, 3 subdural hematomas	Modest (NS) changes in motor tests at 12 months Transplanted cells can survive and integrate anatomically over 10 years	[146]
E9-12 WGE	4	CN + pcPu [B]	Safe; no serious SE	Stabilization or improvement in some neurological indices Prolonged graft survival with development of striatal-like structure	[148]
E7.5-9 WGE	5	2 CN + 3 Pu [B]	Safe procedure	Motor and electrophysiol improvements continue over 6 years	[145, 155]
E8-10 LGE	14	1 CN + 4 Pu [B]	Safe; no serious SEs	Benefit motor, limited neuropsychogic tests	[149, 150]
E8-12 WGE	4	2 CN + 4 Pu [U]	Only SEs related to immunosuppression	Safety only, efficacy not reported	[153]
Porcine LGE	12	2 CN + 4 Pu [U]	Safe; no serious SEs	No change on TFC over 12 months	[161]

CN = caudate nucleus; E = weeks of embryonic age; LLGE = lateral aspect of the lateral ganglionic eminence; NA=not available; NS=not significant; pcPU = postcommissural putamen; Pu = putamen; SEs = side effects; TFC=total functional capacity; WGE = whole ganglionic eminence; VM = ventral mesencephalon; [B] = bilateral implants; [U] = unilateral implants.

Table 5. Clinical trials in HD patients

6.2. Clinical trials of cell transplantation in Huntington's Disease

On the basis of the encouraging results from animal studies, clinical trials using fetal neural stem cells have been performed since 1990 (Table 5). In the early studies performed in Cuba, Czechoslovakia, and Mexico, implantation protocols and procedures were shown not to cause major complications or overt side effects [143, 144]. Afterward, extensive series of implants proved safety of the procedure in moderate stage of patients with functional and/or radiological improvement [145-148]. The first extensive series of fetal striatal tissue implantation was performed in Los Angeles, USA. The procedure was safe [149], and the results indicated motor and cognitive improvements in small numbers of patients [150]. Graft survival was identified by magnetic resonance spectroscopy [151]. Another study group in Tampa, Florida, USA, reported no overall improvement in motor function despite absence of immune rejection [146]. However, when patients with the procedure-dependent side effect (i.e., SDH) were excluded from the result analysis, motor scores were significantly improved after transplantation, and the postmortem brain analysis in one patient indicated healthy surviving grafts 18 months after transplantation and good differentiation into mature striatal-like tissue containing all striatal cell phenotypes [152]. Although three of the patients developed subdural hematoma after the surgery, this side effect could be avoided by careful selection of patients, i.e., patients without significant brain atrophy, for operation. Additional safety evaluation trial involving the United Kingdom arm of the European network for striatal transplantation ("NEST-UK") resulted in no serious side effect related to the operation [153]. This group also suggested a comprehensive, logical yet pragmatic screening program for future neural transplantation [154]. Meanwhile, the efficacy of fetal striatal cell transplantation was shown in another study group, in which four patients with HD have the stabilization or improvement of motor indices after fetal neuroblast implantation, and the graft has survived and developed striatal-like structures in the host brain [148].

Another large-scale study has been going on in a multicenter trial in France (NCT00190450).

This study group is the first to undertake their trial in accordance with the standardized core assessment protocol for intracerebral transplantation in Huntington's disease (CAPIT-HD), and detailed reports of transplantation in the first five patients have been published [145, 147, 155, 156]. In these series, three of the five patients showed motor and cognitive improvements two years after intracerebral fetal neural grafts, which were correlated with recovery of brain metabolic activity in grafted striatal areas and connected regions of the cerebral cortex, measured by fluorodeoxyglucose-positron emission tomography. Restoration of the lost sensory-evoked potentials is also noted. Furthermore, six years after transplantation, clinical improvements became plateaued and then fade off variably within 4–6 years while maintaining stable cognitive function. Cerebral metabolism has also deteriorated progressively, sparing the benefits in the frontal cortex and at the precise location of the grafts. This feature suggested that fetal neural transplantation provides a period of several years of improvement and stability, but not a permanent cure, and strategies of neuroprotection should be developed further [145]. Although it is still controversial, a recent study raised a possibility that the transplanted fetal striatal tissue can undergo disease-like neuronal degeneration after a decade of implantation [157]. Suboptimal long-term graft survival might be caused by the allograft immunoreactivity, microglial responses, and cell-to-cell neurotoxicity [158]. Recently, donor-specific anti-HLA antibodies were detected in six out of 16 patients with HD who received human fetal striatal transplants [159]. These results underline the importance of careful approach for developing cell-based therapy in HD.

Apart from using human fetal striatal tissue, there was a clinical trial using porcine-derived striatal xenografts. In this case, the transplanted patients were treated with cyclosporin or a monoclonal antibody directed against surface major histocompatibility complex I molecules [160]. However, the surviving grafts were not detectable on MRI, and the treatment gave rise to no functional benefit for the patients [161]. In the case of xenografts, fully effective immunosuppression strategies should be resolved.

In order to use neural stem cells for clinical trials, several aborted human fetal tissues are required. Whether derived from elective or spontaneous abortion, there are sensitive ethical and social issues associated with using human fetal tissues for transplantation. There are also difficulties with accurate staging and collection (subject to appropriate ethical approval) and storage of fetal tissues [162-164]. Therefore, in most cases, they would never meet the levels of standardization and quality control required.

The delivery of trophic factors by genetically modified cells into striatum of patient has become another therapeutic approach in HD. A variety of growth factors, including the neurotrophins, fibroblast growth factor (FGF), ciliary neurotrophic factor (CNTF), glia-derived neurotrophic factor (GDNF), and brain-derived neurotrophic factor (BDNF), have been found to promote survival of striatal neurons in culture [117, 118, 165]. Direct injection of various trophic factors incorporated into viral vectors has problems of standardization, regulation, and longevity of treatment. Instead, implantation of engineered cells to express the desired transgene has been suggested [166]. Intracerebral administration of a device formed by a semipermeable membrane encapsulating a BHK cell line engineered to synthesize CNTF has been tried. No sign of CNTF-induced toxicity was observed, while improvements in electrophysiological results were observed. However, depression was observed in three out of six HD patients, and heterogeneous cell survival in the device hindered further clinical trials [166, 167]. Recently, mesenchymal stem cell engineered to secrete BDNF has been suggested, due to its clinically applicable characteristics [116].

7. Future directions of Huntington's Disease research and clinical applications

Therapeutic approaches using NSCs and other stem cell products for CNS diseases fall into two broad categories: (i) regenerative/cell replacement to promote host tissue repair mechanisms and/or to replace missing or damaged cells and (ii) therapeutic delivery of macromolecules (enzymes, cytokines, neurotrophins, drugs, etc.) for neuroprotection and/or stimulation of repair. However, because HD is a progressive neurodegenerative disease and stem cells might replace and protect only striatal neurons with limited capacity, stem cell therapy as

http://dx.doi.org/10.5772/60618

means to stop disease progression might be insufficient [168]. As such, future directions of cell therapy in HD should move beyond the replacement of lost neurons. To date, clinical trials have been undertaken with fetal donor tissue in the striatum in HD. For therapeutic efficacy, reconstitution of a functional dynamic information-processing circuit without ectopic connections using transplanted stem cells is necessary. Another major challenge is to achieve controlled differentiation of embryonic stem cells or induced pluripotent stem cells into specific neuronal phenotypes, such as medium spiny neurons in the absence of aberrant growth or tumor formation. Furthermore, novel approaches to provide therapeutic molecules for neuroprotection should also be tried and verified.

Acknowledgements

This work was supported by a grant of the Basic Science Research Program through the National Research Foundation of Korea (NRF) funded by the Ministry of Education, Science, and Technology (2012R1A1A2006827), Republic of Korea.

Author details

Hyun Sook Kim[1,2] and Jihwan Song[2*]

*Address all correspondence to: jsong@cha.ac.kr

1 Department of Neurology, CHA Bundang Medical Center, CHA University, South Korea

2 CHA Stem Cell Institute, CHA University, South Korea

References

[1] Andrade MA, Bork P: HEAT repeats in the Huntington's disease protein. Nature genetics. 1995;11(2):115–6.

[2] Sharp AH, Loev SJ, Schilling G, Li SH, Li XJ, Bao J, et al: Widespread expression of Huntington's disease gene (IT15) protein product. Neuron. 1995;14(5):1065–74.

[3] DiFiglia M, Sapp E, Chase K, Schwarz C, Meloni A, Young C, et al: Huntingtin is a cytoplasmic protein associated with vesicles in human and rat brain neurons. Neuron. 1995;14(5):1075–81.

[4] De Rooij KE, Dorsman JC, Smoor MA, Den Dunnen JT, Van Ommen GJ: Subcellular localization of the Huntington's disease gene product in cell lines by immunofluores-

cence and biochemical subcellular fractionation. Human Molecular Genetics. 1996;5(8):1093–9.

[5] Savas JN, Ma B, Deinhardt K, Culver BP, Restituito S, Wu L, et al: A role for huntington disease protein in dendritic RNA granules. Journal of Biological Chemistry. 2010;285(17):13142–53.

[6] Sugars KL, Rubinsztein DC: Transcriptional abnormalities in Huntington disease. Trends in Genetics. 2003;19(5):233–8.

[7] Buckley NJ, Johnson R, Zuccato C, Bithell A, Cattaneo E: The role of REST in transcriptional and epigenetic dysregulation in Huntington's disease. Neurobiology of Disease. 2010;39(1):28–39.

[8] Cattaneo E, Zuccato C, Tartari M: Normal huntingtin function: an alternative approach to Huntington's disease. Nature Reviews Neuroscience. 2005;6(12):919–30.

[9] Zuccato C, Liber D, Ramos C, Tarditi A, Rigamonti D, Tartari M, et al: Progressive loss of BDNF in a mouse model of Huntington's disease and rescue by BDNF delivery. Pharmacological Research. 2005;52(2):133–9.

[10] Zeitlin S, Liu JP, Chapman DL, Papaioannou VE, Efstratiadis A: Increased apoptosis and early embryonic lethality in mice nullizygous for the Huntington's disease gene homologue. Nature Genetics. 1995;11(2):155–63.

[11] Nagai Y, Inui T, Popiel HA, Fujikake N, Hasegawa K, Urade Y, et al: A toxic monomeric conformer of the polyglutamine protein. Nature Structural and Molecular Biology. 2007;14(4):332–40.

[12] Zuccato CMV, Cattaneo E: Molecular mechanisms and potential therapeutic targets in Huntington's disease. Physiological Reviews. 2010;90:905–81.

[13] Ross CA, Tabrizi SJ: Huntington's disease: from molecular pathogenesis to clinical treatment. Lancet Neurology. 2011;10(1):83–98.

[14] Leegwater-Kim J, Cha JH: The paradigm of Huntington's disease: therapeutic opportunities in neurodegeneration. NeuroRx: The Journal of the American Society for Experimental NeuroTherapeutics. 2004;1(1):128–38.

[15] Shao J, Diamond MI: Polyglutamine diseases: emerging concepts in pathogenesis and therapy. Human Molecular Genetics. 2007;16(Spec No. 2):R115–23.

[16] Tobin AJ, Signer ER: Huntington's disease: the challenge for cell biologists. Trends in Cell Biology. 2000;10(12):531–6.

[17] Pringsheim T, Wiltshire K, Day L, Dykeman J, Steeves T, Jette N: The incidence and prevalence of Huntington's disease: a systematic review and meta-analysis. Movement Disorders. 2012;27(9):1083–91.

[18] Gusella JF, MacDonald ME, Lee JM: Genetic modifiers of Huntington's disease. Movement Disorders. 2014;29(11):1359–65.

[19] Langbehn DR, Brinkman RR, Falush D, Paulsen JS, Hayden MR, International Huntington's Disease Collaborative G: A new model for prediction of the age of onset and penetrance for Huntington's disease based on CAG length. Clinical Genetics. 2004;65(4):267–77.

[20] Wexler NS, Lorimer J, Porter J, Gomez F, Moskowitz C, Shackell E, et al: Venezuelan kindreds reveal that genetic and environmental factors modulate Huntington's disease age of onset. Proceedings of the National Academy of Sciences of the United States of America. 2004;101(10):3498–503.

[21] Lee JM, Ramos EM, Lee JH, Gillis T, Mysore JS, Hayden MR, et al: CAG repeat expansion in Huntington disease determines age at onset in a fully dominant fashion. Neurology. 2012;78:690–5.

[22] Huntington G: On chorea. Medical and Surgical Reporter. 1872;26:317–21.

[23] van der Burg JM, Bjorkqvist M, Brundin P: Beyond the brain: widespread pathology in Huntington's disease. Lancet Neurology. 2009;8(8):765–74.

[24] Halliday GM, McRitchie DA, Macdonald V, Double KL, Trent RJ, McCusker E: Regional specificity of brain atrophy in Huntington's disease. Experimental Neurology. 1998;154(2):663–72.

[25] Subramaniam S, Sixt KM, Barrow R, Snyder SH: Rhes, a striatal specific protein, mediates mutant-huntingtin cytotoxicity. Science. 2009;324(5932):1327–30.

[26] Rockabrand E, Slepko N, Pantalone A, Nukala VN, Kazantsev A, Marsh JL, et al: The first 17 amino acids of Huntingtin modulate its sub-cellular localization, aggregation and effects on calcium homeostasis. Human Molecular Genetics. 2007;16(1):61–77.

[27] Gu X, Greiner ER, Mishra R, Kodali R, Osmand A, Finkbeiner S, et al: Serines 13 and 16 are critical determinants of full-length human mutant huntingtin induced disease pathogenesis in HD mice. Neuron. 2009;64(6):828–40.

[28] Yanai A, Huang K, Kang R, Singaraja RR, Arstikaitis P, Gan L, et al: Palmitoylation of huntingtin by HIP14 is essential for its trafficking and function. Nature Neuroscience. 2006;9(6):824–31.

[29] Cowan CM, Raymond LA: Selective neuronal degeneration in Huntington's disease. Current Topics in Developmental Biology. 2006;75:25–71.

[30] Shin JY, Fang ZH, Yu ZX, Wang CE, Li SH, Li XJ: Expression of mutant huntingtin in glial cells contributes to neuronal excitotoxicity. Journal of Cell Biology. 2005;171(6): 1001–12.

[31] Juopperi TA, Kim WR, Chiang CH, Yu H, Margolis RL, Ross CA, et al: Astrocytes generated from patient induced pluripotent stem cells recapitulate features of Huntington's disease patient cells. Molecular Brain. 2012;5(1):17.

[32] Bradford J, Shin JY, Roberts M, Wang CE, Li XJ, Li S: Expression of mutant huntingtin in mouse brain astrocytes causes age-dependent neurological symptoms. Proceedings of the National Academy of Sciences of the United States of America. 2009;106(52):22480–5.

[33] Pavese N, Gerhard A, Tai YF, Ho AK, Turkheimer F, Barker RA, et al: Microglial activation correlates with severity in Huntington disease: a clinical and PET study. Neurology. 2006;66(11):1638–43.

[34] Tai YF, Pavese N, Gerhard A, Tabrizi SJ, Barker RA, Brooks DJ, et al: Microglial activation in presymptomatic Huntington's disease gene carriers. Brain. 2007;130(Pt 7): 1759–66.

[35] Kenney C, Jankovic J: Tetrabenazine in the treatment of hyperkinetic movement disorders. Expert Review of Neurotherapeutics. 2006;6(1):7–17.

[36] Fernandez HH, Friedman JH, Grace J, Beason-Hazen S: Donepezil for Huntington's disease. Movement Disorders. 2000;15(1):173–6.

[37] de Tommaso M, Difruscolo O, Sciruicchio V, Specchio N, Livrea P: Two years' follow-up of rivastigmine treatment in Huntington disease. Clinical Neuropharmacology. 2007;30(1):43–6.

[38] Nance MA, Sanders G: Characteristics of individuals with Huntington disease in long-term care. Movement Disorders. 1996;11(5):542–8.

[39] Myers RH, Sax DS, Koroshetz WJ, Mastromauro C, Cupples LA, Kiely DK, et al: Factors associated with slow progression in Huntington's disease. Archives of Neurology. 1991;48(8):800–4.

[40] Benn CL, Luthi-Carter R, Kuhn A, Sadri-Vakili G, Blankson KL, Dalai SC, et al: Environmental enrichment reduces neuronal intranuclear inclusion load but has no effect on messenger RNA expression in a mouse model of Huntington disease. Journal of Neuropathology and Experimental Neurology. 2010;69(8):817–27.

[41] van Dellen A, Blakemore C, Deacon R, York D, Hannan AJ: Delaying the onset of Huntington's in mice. Nature. 2000;404(6779):721–2.

[42] Landwehrmeyer GB, Dubois B, de Yebenes JG, Kremer B, Gaus W, Kraus PH, et al: Riluzole in Huntington's disease: a 3-year, randomized controlled study. Annals of Neurology. 2007;62(3):262–72.

[43] Peyser CE, Folstein M, Chase GA, Starkstein S, Brandt J, Cockrell JR, et al: Trial of d-alpha-tocopherol in Huntington's disease American Journal of Psychiatry. 1995;152(12):1771–5.

[44] Huntington Study Group: A randomized, placebo-controlled trial of coenzyme Q10 and remacemide in Huntington's disease. Neurology. 2001;57(3):397–404.

[45] Huntington Study Group: A futility study of minocycline in Huntington's disease. Movement Disorders. 2010;25(13):2219–24.

[46] Mestre T, Ferreira J, Coelho MM, Rosa M, Sampaio C: Therapeutic interventions for disease progression in Huntington's disease. The Cochrane Database of Systematic Reviews. 2009(3):CD006455.

[47] Wild EJ, Tabrizi SJ: Targets for future clinical trials in Huntington's disease: what's in the pipeline? Movement Disorders. 2014;29(11):1434–45.

[48] Ferrante RJ: Mouse models of Huntington's disease and methodological considerations for therapeutic trials. Biochimica et Biophysica Acta. 2009;1792(6):506–20.

[49] Schwarcz R, Coyle JT: Striatal lesions with kainic acid: neurochemical characteristics. Brain Research. 1977;127(2):235–49.

[50] Schwarcz R, Whetsell WO, Jr., Mangano RM: Quinolinic acid: an endogenous metabolite that produces axon-sparing lesions in rat brain. Science. 1983;219(4582):316–8.

[51] Beal MF, Brouillet E, Jenkins BG, Ferrante RJ, Kowall NW, Miller JM, et al: Neurochemical and histologic characterization of striatal excitotoxic lesions produced by the mitochondrial toxin 3-nitropropionic acid. Journal of Neuroscience. 1993;13(10): 4181–92.

[52] Brouillet E, Conde F, Beal MF, Hantraye P: Replicating Huntington's disease phenotype in experimental animals. Progress in Neurobiology. 1999;59(5):427–68.

[53] Dunnett SB, Rosser AE: Cell therapy in Huntington's disease. NeuroRx. 2004;1(4): 394–405.

[54] Cepeda C, Cummings DM, Andre VM, Holley SM, Levine MS: Genetic mouse models of Huntington's disease: focus on electrophysiological mechanisms. ASN Neuro. 2010;2(2):e00033.

[55] Pouladi MA, Morton AJ, Hayden MR: Choosing an animal model for the study of Huntington's disease. Nature Reviews Neuroscience. 2013;14(10):708–21.

[56] Mangiarini L, Sathasivam K, Seller M, Cozens B, Harper A, Hetherington C, et al: Exon 1 of the HD gene with an expanded CAG repeat is sufficient to cause a progressive neurological phenotype in transgenic mice. Cell. 1996;87(3):493–506.

[57] Naver B, Stub C, Moller M, Fenger K, Hansen AK, Hasholt L, et al: Molecular and behavioral analysis of the R6/1 Huntington's disease transgenic mouse. Neuroscience. 2003;122(4):1049–57.

[58] Laforet GA, Sapp E, Chase K, McIntyre C, Boyce FM, Campbell M, et al: Changes in cortical and striatal neurons predict behavioral and electrophysiological abnormali-

ties in a transgenic murine model of Huntington's disease. Journal of Neuroscience. 2001;21(23):9112–23.

[59] Schilling G, Becher MW, Sharp AH, Jinnah HA, Duan K, Kotzuk JA, et al: Intranuclear inclusions and neuritic aggregates in transgenic mice expressing a mutant N-terminal fragment of huntingtin. Human Molecular Genetics. 1999;8(3):397–407.

[60] von Horsten S, Schmitt I, Nguyen HP, Holzmann C, Schmidt T, Walther T, et al: Transgenic rat model of Huntington's disease. Human Molecular Genetics. 2003;12(6):617–24.

[61] Hodgson JG, Smith DJ, McCutcheon K, Koide HB, Nishiyama K, Dinulos MB, et al: Human huntingtin derived from YAC transgenes compensates for loss of murine huntingtin by rescue of the embryonic lethal phenotype. Human Molecular Genetics. 1996;5(12):1875–85.

[62] Slow EJ, van Raamsdonk J, Rogers D, Coleman SH, Graham RK, Deng Y, et al: Selective striatal neuronal loss in a YAC128 mouse model of Huntington disease. Human Molecular Genetics. 2003;12(13):1555–67.

[63] Reddy PH, Williams M, Charles V, Garrett L, Pike-Buchanan L, Whetsell WO, Jr., et al: Behavioural abnormalities and selective neuronal loss in HD transgenic mice expressing mutated full-length HD cDNA. Nature Genetics. 1998;20(2):198–202.

[64] Gray M, Shirasaki DI, Cepeda C, Andre VM, Wilburn B, Lu XH, et al: Full-length human mutant huntingtin with a stable polyglutamine repeat can elicit progressive and selective neuropathogenesis in BACHD mice. Journal of Neuroscience. 2008;28(24): 6182–95.

[65] Southwell AL, Warby SC, Carroll JB, Doty CN, Skotte NH, Zhang W, et al: A fully humanized transgenic mouse model of Huntington disease. Human Molecular Genetics. 2013;22(1):18–34.

[66] Wheeler VC, Auerbach W, White JK, Srinidhi J, Auerbach A, Ryan A, et al: Length-dependent gametic CAG repeat instability in the Huntington's disease knock-in mouse. Human Molecular Genetics. 1999;8(1):115–22.

[67] Wheeler VC, Gutekunst CA, Vrbanac V, Lebel LA, Schilling G, Hersch S, et al: Early phenotypes that presage late-onset neurodegenerative disease allow testing of modifiers in Hdh CAG knock-in mice. Human Molecular Genetics. 2002;11(6):633–40.

[68] Wheeler VC, White JK, Gutekunst CA, Vrbanac V, Weaver M, Li XJ, et al: Long glutamine tracts cause nuclear localization of a novel form of huntingtin in medium spiny striatal neurons in HdhQ92 and HdhQ111 knock-in mice. Human Molecular Genetics. 2000;9(4):503–13.

[69] Menalled LB, Sison JD, Dragatsis I, Zeitlin S, Chesselet MF: Time course of early motor and neuropathological anomalies in a knock-in mouse model of Huntington's disease with 140 CAG repeats. Journal of Comparative Neurology. 2003;465(1):11–26.

[70] Lin CH, Tallaksen-Greene S, Chien WM, Cearley JA, Jackson WS, Crouse AB, et al: Neurological abnormalities in a knock-in mouse model of Huntington's disease. Human Molecular Genetics. 2001;10(2):137–44.

[71] Tabrizi SJ, Workman J, Hart PE, Mangiarini L, Mahal A, Bates G, et al: Mitochondrial dysfunction and free radical damage in the Huntington R6/2 transgenic mouse. Annals of Neurology. 2000;47(1):80–6.

[72] Cha JH, Kosinski CM, Kerner JA, Alsdorf SA, Mangiarini L, Davies SW, et al: Altered brain neurotransmitter receptors in transgenic mice expressing a portion of an abnormal human huntington disease gene. Proceedings of the National Academy of Sciences of the United States of America. 1998;95(11):6480–5.

[73] Murphy KP, Carter RJ, Lione LA, Mangiarini L, Mahal A, Bates GP, et al: Abnormal synaptic plasticity and impaired spatial cognition in mice transgenic for exon 1 of the human Huntington's disease mutation. Journal of Neuroscience. 2000;20(13):5115–23.

[74] Cummings DM, Milnerwood AJ, Dallerac GM, Vatsavayai SC, Hirst MC, Murphy KP: Abnormal cortical synaptic plasticity in a mouse model of Huntington's disease. Brain Research Bulletin. 2007;72(2–3):103–7.

[75] Menalled LB, Chesselet MF: Mouse models of Huntington's disease. Trends in Pharmacological Sciences. 2002;23(1):32–9.

[76] Hodgson JG, Agopyan N, Gutekunst CA, Leavitt BR, LePiane F, Singaraja R, et al: A YAC mouse model for Huntington's disease with full-length mutant huntingtin, cytoplasmic toxicity, and selective striatal neurodegeneration. Neuron. 1999;23(1):181–92.

[77] Van Raamsdonk JM, Warby SC, Hayden MR: Selective degeneration in YAC mouse models of Huntington disease. Brain Research Bulletin. 2007;72(2–3):124–31.

[78] Yang XW, Gray M: Mouse Models for Validating Preclinical Candidates for Huntington's Disease. In: Lo DC, Hughes RE, editors. Neurobiology of Huntington's Disease: Applications to Drug Discovery. Frontiers in Neuroscience. Boca Raton (FL): CRC Press; 2011. Chapter 7. Available from: http://www.ncbi.nlm.nih.gov/books/NBK56001/

[79] Burns LH, Pakzaban P, Deacon TW, Brownell AL, Tatter SB, Jenkins BG, et al: Selective putaminal excitotoxic lesions in non-human primates model the movement disorder of Huntington disease. Neuroscience. 1995;64(4):1007–17.

[80] Yang SH, Cheng PH, Banta H, Piotrowska-Nitsche K, Yang JJ, Cheng EC, et al: Towards a transgenic model of Huntington's disease in a non-human primate. Nature. 2008;453(7197):921–4.

[81] Yang D, Wang CE, Zhao B, Li W, Ouyang Z, Liu Z, et al: Expression of Huntington's disease protein results in apoptotic neurons in the brains of cloned transgenic pigs. Human Molecular Genetics. 2010;19(20):3983–94.

[82] Jacobsen JC, Bawden CS, Rudiger SR, McLaughlan CJ, Reid SJ, Waldvogel HJ, et al: An ovine transgenic Huntington's disease model. Human Molecular Genetics. 2010;19(10):1873–82.

[83] Kocerha J, Liu Y, Willoughby D, Chidamparam K, Benito J, Nelson K, et al: Longitudinal transcriptomic dysregulation in the peripheral blood of transgenic Huntington's disease monkeys. BMC Neuroscience. 2013;14:88.

[84] Huntington's Disease Sheep Collaborative Research Group, Reid SJ, Patassini S, Handley RR, Rudiger SR, McLaughlan CJ, et al: Further molecular characterisation of the OVT73 transgenic sheep model of Huntington's disease identifies cortical aggregates. Journal of Huntington's Disease. 2013;2(3):279–95.

[85] Baxa M, Hruska-Plochan M, Juhas S, Vodicka P, Pavlok A, Juhasova J, et al: A transgenic minipig model of Huntington's disease. Journal of Huntington's Disease. 2013;2(1):47–68.

[86] Lindvall O, Kokaia Z, Martinez-Serrano A: Stem cell therapy for human neurodegenerative disorders-how to make it work. Nature Medicine. 2004;10 Suppl:S42–50.

[87] Bjorklund A, Lindvall O: Cell replacement therapies for central nervous system disorders. Nature Neuroscience. 2000;3(6):537–44.

[88] Dunnett SB, Rosser AE: Stem cell transplantation for Huntington's disease. Experimental Neurology. 2007;203(2):279–92.

[89] Armstrong RJ, Svendsen CN: Neural stem cells: from cell biology to cell replacement. Cell Transplantation. 2000;9(2):139–52.

[90] Armstrong RJ, Watts C, Svendsen CN, Dunnett SB, Rosser AE: Survival, neuronal differentiation, and fiber outgrowth of propagated human neural precursor grafts in an animal model of Huntington's disease. Cell Transplantation. 2000;9(1):55–64.

[91] Lillien L: Neural progenitors and stem cells: mechanisms of progenitor heterogeneity. Current Opinion in Neurobiology. 1998;8(1):37–44.

[92] McBride JL, Behrstock SP, Chen EY, Jakel RJ, Siegel I, Svendsen CN, et al: Human neural stem cell transplants improve motor function in a rat model of Huntington's disease. Journal of Comparative Neurology. 2004;475(2):211–9.

[93] Lee ST, Chu K, Park JE, Lee K, Kang L, Kim SU, et al: Intravenous administration of human neural stem cells induces functional recovery in Huntington's disease rat model. Neuroscience Research. 2005;52(3):243–9.

[94] Johann V, Schiefer J, Sass C, Mey J, Brook G, Kruttgen A, et al: Time of transplantation and cell preparation determine neural stem cell survival in a mouse model of Huntington's disease. Experimental Brain Research. 2007;177(4):458–70.

[95] Kordower JH, Chen EY, Winkler C, Fricker R, Charles V, Messing A, et al: Grafts of EGF-responsive neural stem cells derived from GFAP-hNGF transgenic mice: trophic

and tropic effects in a rodent model of Huntington's disease. Journal of Comparative Neurology. 1997;387(1):96–113.

[96] Chung S, Shin BS, Hwang M, Lardaro T, Kang UJ, Isacson O, et al: Neural precursors derived from embryonic stem cells, but not those from fetal ventral mesencephalon, maintain the potential to differentiate into dopaminergic neurons after expansion in vitro. Stem Cells. 2006;24(6):1583–93.

[97] El Akabawy G, Rattray I, Johansson SM, Gale R, Bates G, Modo M: Implantation of undifferentiated and predifferentiated human neural stem cells in the R6/2 transgenic mouse model of Huntington's disease. BMC Neuroscience. 2012;13(1):97.

[98] Reubinoff BE, Pera MF, Fong CY, Trounson A, Bongso A: Embryonic stem cell lines from human blastocysts: somatic differentiation in vitro. Nature Biotechnology. 2000;18(4):399–404.

[99] Zhang SC, Wernig M, Duncan ID, Brustle O, Thomson JA: In vitro differentiation of transplantable neural precursors from human embryonic stem cells. Nature Biotechnology. 2001;19(12):1129–33.

[100] Dinsmore J, Ratliff J, Deacon T, Pakzaban P, Jacoby D, Galpern W, et al: Embryonic stem cells differentiated in vitro as a novel source of cells for transplantation. Cell transplantation. 1996;5(2):131–43.

[101] Gage FH. Mammalian neural stem cells. Science. 2000;287(5457):1433–8.

[102] Vazey EM, Dottori M, Jamshidi P, Tomas D, Pera MF, Horne M, et al: Comparison of transplant efficiency between spontaneously derived and noggin-primed human embryonic stem cell neural precursors in the quinolinic acid rat model of Huntington's disease. Cell Transplantation. 2010;19(8):1055–62.

[103] Shin E, Palmer MJ, Li M, Fricker RA: GABAergic neurons from mouse embryonic stem cells possess functional properties of striatal neurons in vitro, and develop into striatal neurons in vivo in a mouse model of Huntington's disease. Stem Cell Reviews. 2012;8(2):513–31.

[104] Aubry L, Bugi A, Lefort N, Rousseau F, Peschanski M, Perrier AL: Striatal progenitors derived from human ES cells mature into DARPP32 neurons in vitro and in quinolinic acid-lesioned rats. Proceedings of the National Academy of Sciences of the United States of America. 2008;105(43):16707–12.

[105] Martino G, Pluchino S: The therapeutic potential of neural stem cells. Nature Reviews Neuroscience. 2006;7(5):395–406.

[106] Amin EM, Reza BA, Morteza BR, Maryam MM, Ali M, Zeinab N: Microanatomical evidences for potential of mesenchymal stem cells in amelioration of striatal degeneration. Neurological Research. 2008;30(10):1086–90.

[107] Sadan O, Shemesh N, Barzilay R, Bahat-Stromza M, Melamed E, Cohen Y, et al: Migration of neurotrophic factors-secreting mesenchymal stem cells toward a quinolinic

acid lesion as viewed by magnetic resonance imaging. Stem Cells. 2008;26(10):2542–51.

[108] Edalatmanesh MA, Matin MM, Neshati Z, Bahrami AR, Kheirabadi M: Systemic transplantation of mesenchymal stem cells can reduce cognitive and motor deficits in rats with unilateral lesions of the neostriatum. Neurological Research. 2010;32(2): 166–72.

[109] Lee ST, Chu K, Jung KH, Im WS, Park JE, Lim HC, et al: Slowed progression in models of Huntington disease by adipose stem cell transplantation. Annals of Neurology. 2009;66(5):671–81.

[110] Im W, Lee ST, Park JE, Oh HJ, Shim J, Lim J, et al: Transplantation of patient-derived adipose stem cells in YAC128 Huntington's disease transgenic mice. PLoS Curr. 2010;2.

[111] Long Y, Yang KY: Bone marrow derived cells for brain repair: recent findings and current controversies. Current Molecular Medicine. 2003;3(8):719–25.

[112] Rossignol J, Boyer C, Leveque X, Fink KD, Thinard R, Blanchard F, et al: Mesenchymal stem cell transplantation and DMEM administration in a 3NP rat model of Huntington's disease: morphological and behavioral outcomes. Behavioural Brain Research. 2011;217(2):369–78.

[113] Lin YT, Chern Y, Shen CK, Wen HL, Chang YC, Li H, et al: Human mesenchymal stem cells prolong survival and ameliorate motor deficit through trophic support in Huntington's disease mouse models. PloS One. 2011;6(8):e22924.

[114] Jiang Y, Lv H, Huang S, Tan H, Zhang Y, Li H: Bone marrow mesenchymal stem cells can improve the motor function of a Huntington's disease rat model. Neurological Research. 2011;33(3):331–7.

[115] Moraes L, Vasconcelos-dos-Santos A, Santana FC, Godoy MA, Rosado-de-Castro PH, Jasmin, et al: Neuroprotective effects and magnetic resonance imaging of mesenchymal stem cells labeled with SPION in a rat model of Huntington's disease. Stem Cell Research. 2012;9(2):143–55.

[116] Olson SD, Pollock K, Kambal A, Cary W, Mitchell GM, Tempkin J, et al: Genetically engineered mesenchymal stem cells as a proposed therapeutic for Huntington's disease. Molecular Neurobiology. 2012;45(1):87–98.

[117] Ventimiglia R, Mather PE, Jones BE, Lindsay RM: The neurotrophins BDNF, NT-3 and NT-4/5 promote survival and morphological and biochemical differentiation of striatal neurons in vitro. European journal of Neuroscience. 1995;7(2):213–22.

[118] Nakao N, Odin P, Lindvall O, Brundin P: Differential trophic effects of basic fibroblast growth factor, insulin-like growth factor-1, and neurotrophin-3 on striatal neurons in culture. Experimental Neurology. 1996;138(1):144–57.

[119] Sadan O, Shemesh N, Barzilay R, Dadon-Nahum M, Blumenfeld-Katzir T, Assaf Y, et al: Mesenchymal stem cells induced to secrete neurotrophic factors attenuate quinolinic acid toxicity: a potential therapy for Huntington's disease. Experimental Neurology. 2012;234(2):417–27.

[120] Dey ND, Bombard MC, Roland BP, Davidson S, Lu M, Rossignol J, et al: Genetically engineered mesenchymal stem cells reduce behavioral deficits in the YAC 128 mouse model of Huntington's disease. Behavioural Brain Research. 2010;214(2):193–200.

[121] Im W, Ban J, Lim J, Lee M, Lee ST, Chu K, et al: Extracts of adipose derived stem cells slows progression in the R6/2 model of Huntington's disease. PloS One. 2013;8(4):e59438.

[122] Park IH, Arora N, Huo H, Maherali N, Ahfeldt T, Shimamura A, et al: Disease-specific induced pluripotent stem cells. Cell. 2008;134(5):877–86.

[123] Takahashi K, Tanabe K, Ohnuki M, Narita M, Ichisaka T, Tomoda K, et al: Induction of pluripotent stem cells from adult human fibroblasts by defined factors. Cell. 2007;131(5):861–72.

[124] Perrier A, Peschanski M: How can human pluripotent stem cells help decipher and cure Huntington's disease? Cell Stem Cell. 2012;11(2):153–61.

[125] Yu J, Hu K, Smuga-Otto K, Tian S, Stewart R, Slukvin, II, et al: Human induced pluripotent stem cells free of vector and transgene sequences. Science. 2009;324(5928):797–801.

[126] HD iPSC Consortium: Induced pluripotent stem cells from patients with Huntington's disease show CAG-repeat-expansion-associated phenotypes. Cell Stem Cell. 2012;11(2):264–78.

[127] Jeon I, Lee N, Li JY, Park IH, Park KS, Moon J, et al: Neuronal properties, in vivo effects, and pathology of a Huntington's disease patient-derived induced pluripotent stem cells. Stem Cells. 2012;30(9):2054–62.

[128] Hockemeyer D, Wang H, Kiani S, Lai CS, Gao Q, Cassady JP, et al: Genetic engineering of human pluripotent cells using TALE nucleases. Nature Biotechnology. 2011;29(8):731–4.

[129] Smith C, Abalde-Atristain L, He C, Brodsky BR, Braunstein EM, Chaudhari P, et al: Efficient and allele-specific genome editing of disease loci in human iPSCs. Molecular Therapy. 2014. doi: 10.1038/mt.2014.226.

[130] An MC, Zhang N, Scott G, Montoro D, Wittkop T, Mooney S, et al: Genetic correction of Huntington's disease phenotypes in induced pluripotent stem cells. Cell Stem Cell. 2012;11(2):253–63.

[131] Dobrossy MD, Dunnett SB: Striatal grafts alleviate deficits in response execution in a lateralised reaction time task. Brain Research Bulletin. 1998;47(6):585–93.

[132] Dunnett SB, Carter RJ, Watts C, Torres EM, Mahal A, Mangiarini L, et al: Striatal transplantation in a transgenic mouse model of Huntington's disease. Experimental Neurology. 1998;154(1):31–40.

[133] van Dellen A, Deacon R, York D, Blakemore C, Hannan AJ: Anterior cingulate cortical transplantation in transgenic Huntington's disease mice. Brain Research Bulletin. 2001;56(3–4):313–8.

[134] Wictorin K: Anatomy and connectivity of intrastriatal striatal transplants. Progress in Neurobiology. 1992;38(6):611–39.

[135] Xu ZC, Wilson CJ, Emson PC: Synaptic potentials evoked in spiny neurons in rat neostriatal grafts by cortical and thalamic stimulation. Journal of Neurophysiology. 1991;65(3):477–93.

[136] Nakao N, Ogura M, Nakai K, Itakura T: Embryonic striatal grafts restore neuronal activity of the globus pallidus in a rodent model of Huntington's disease. Neuroscience. 1999;88(2):469–77.

[137] Sirinathsinghji DJ, Heavens RP, Torres EM, Dunnett SB: Cholecystokinin-dependent regulation of host dopamine inputs to striatal grafts. Neuroscience. 1993;53(3):651–63.

[138] Campbell K, Kalen P, Wictorin K, Lundberg C, Mandel RJ, Bjorklund A: Characterization of GABA release from intrastriatal striatal transplants: dependence on host-derived afferents. Neuroscience. 1993;53(2):403–15.

[139] Sirinathsinghji DJ, Dunnett SB, Isacson O, Clarke DJ, Kendrick K, Bjorklund A: Striatal grafts in rats with unilateral neostriatal lesions – II. In vivo monitoring of GABA release in globus pallidus and substantia nigra. Neuroscience. 1988;24(3):803–11.

[140] Ryu JK, Kim J, Cho SJ, Hatori K, Nagai A, Choi HB, et al: Proactive transplantation of human neural stem cells prevents degeneration of striatal neurons in a rat model of Huntington disease. Neurobiology of Disease. 2004;16(1):68–77.

[141] Kelly CM, Dunnett SB, Rosser AE. Medium spiny neurons for transplantation in Huntington's disease. Biochemical Society Transactions. 2009;37(Pt 1):323–8.

[142] Ma L, Hu B, Liu Y, Vermilyea SC, Liu H, Gao L, et al: Human embryonic stem cell-derived GABA neurons correct locomotion deficits in quinolinic acid-lesioned mice. Cell Stem Cell. 2012;10(4):455–64.

[143] Sramka M, Rattaj M, Molina H, Vojtassak J, Belan V, Ruzicky E: Stereotactic technique and pathophysiological mechanisms of neurotransplantation in Huntington's chorea. Stereotactic and Functional Neurosurgery. 1992;58(1–4):79–83.

[144] Madrazo I, Franco-Bourland RE, Castrejon H, Cuevas C, Ostrosky-Solis F: Fetal striatal homotransplantation for Huntington's disease: first two case reports. Neurological Research. 1995;17(4):312–5.

[145] Bachoud-Levi AC, Gaura V, Brugieres P, Lefaucheur JP, Boisse MF, Maison P, et al: Effect of fetal neural transplants in patients with Huntington's disease 6 years after surgery: a long-term follow-up study. Lancet Neurology. 2006;5(4):303–9.

[146] Hauser RA, Furtado S, Cimino CR, Delgado H, Eichler S, Schwartz S, et al: Bilateral human fetal striatal transplantation in Huntington's disease. Neurology. 2002;58(5): 687–95.

[147] Gaura V, Bachoud-Levi AC, Ribeiro MJ, Nguyen JP, Frouin V, Baudic S, et al: Striatal neural grafting improves cortical metabolism in Huntington's disease patients. Brain. 2004;127(Pt 1):65–72.

[148] Gallina P, Paganini M, Lombardini L, Mascalchi M, Porfirio B, Gadda D, et al: Human striatal neuroblasts develop and build a striatal-like structure into the brain of Huntington's disease patients after transplantation. Experimental Neurology. 2010;222(1):30–41.

[149] Kopyov OV, Jacques S, Lieberman A, Duma CM, Eagle KS: Safety of intrastriatal neurotransplantation for Huntington's disease patients. Experimental Neurology. 1998;149(1):97–108.

[150] Philpott LM, Kopyov OV, Lee AJ, Jacques S, Duma CM, Caine S, et al: Neuropsychological functioning following fetal striatal transplantation in Huntington's chorea: three case presentations. Cell Transplantation. 1997;6(3):203–12.

[151] Ross BD, Hoang TQ, Bluml S, Dubowitz D, Kopyov OV, Jacques DB, et al: In vivo magnetic resonance spectroscopy of human fetal neural transplants. NMR in Biomedicine. 1999;12(4):221–36.

[152] Freeman TB, Cicchetti F, Hauser RA, Deacon TW, Li XJ, Hersch SM, et al: Transplanted fetal striatum in Huntington's disease: phenotypic development and lack of pathology. Proceedings of the National Academy of Sciences of the United States of America. 2000;97(25):13877–82.

[153] Rosser AE, Barker RA, Harrower T, Watts C, Farrington M, Ho AK, et al: Unilateral transplantation of human primary fetal tissue in four patients with Huntington's disease: NEST-UK safety report ISRCTN no 36485475. Journal of Neurology, Neurosurgery, and Psychiatry. 2002;73(6):678–85.

[154] Farrington M, Wreghitt TG, Lever AM, Dunnett SB, Rosser AE, Barker RA: Neural transplantation in Huntington's disease: the NEST-UK donor tissue microbiological screening program and review of the literature. Cell Transplantation. 2006;15(4):279–94.

[155] Bachoud-Levi A, Bourdet C, Brugieres P, Nguyen JP, Grandmougin T, Haddad B, et al: Safety and tolerability assessment of intrastriatal neural allografts in five patients with Huntington's disease. Experimental Neurology. 2000;161(1):194–202.

[156] Bachoud-Levi AC, Remy P, Nguyen JP, Brugieres P, Lefaucheur JP, Bourdet C, et al: Motor and cognitive improvements in patients with Huntington's disease after neural transplantation. Lancet. 2000;356(9246):1975–9.

[157] Cicchetti F, Saporta S, Hauser RA, Parent M, Saint-Pierre M, Sanberg PR, et al: Neural transplants in patients with Huntington's disease undergo disease-like neuronal degeneration. Proceedings of the National Academy of Sciences of the United States of America. 2009;106(30):12483–8.

[158] Cicchetti F, Soulet D, Freeman TB: Neuronal degeneration in striatal transplants and Huntington's disease: potential mechanisms and clinical implications. Brain. 2011;134(Pt 3):641–52.

[159] Porfirio B, Paganini M, Mazzanti B, Bagnoli S, Bucciantini S, Ghelli E, et al: Donor-specific anti-HLA antibodies in Huntington's disease recipients of human fetal striatal grafts. Cell Transplantation. 2013. doi: 10.3727/096368913X676222.

[160] Pakzaban P, Deacon TW, Burns LH, Dinsmore J, Isacson O: A novel mode of immunoprotection of neural xenotransplants: masking of donor major histocompatibility complex class I enhances transplant survival in the central nervous system. Neuroscience. 1995;65(4):983–96.

[161] Fink JS, Schumacher JM, Ellias SL, Palmer EP, Saint-Hilaire M, Shannon K, et al: Porcine xenografts in Parkinson's disease and Huntington's disease patients: preliminary results. Cell Transplantation. 2000;9(2):273–8.

[162] Rosser AE, Barker RA, Armstrong RJ, Elneil S, Jain M, Hurelbrink CB, et al: Staging and preparation of human fetal striatal tissue for neural transplantation in Huntington's disease. Cell Transplantation. 2003;12(7):679–86.

[163] Nauert GM, Freeman TB: Low-pressure aspiration abortion for obtaining embryonic and early gestational fetal tissue for research purposes. Cell Transplantation. 1994;3(2):147–51.

[164] Hurelbrink CB, Tyers P, Armstrong RJ, Dunnett SB, Barker RA, Rosser AE: Long-term hibernation of human fetal striatal tissue does not adversely affect its differentiation in vitro or graft survival: implications for clinical trials in Huntington's disease. Cell Transplantation. 2003;12(7):687–95.

[165] Petersen A, Brundin P: Effects of ciliary neurotrophic factor on excitotoxicity and calcium-ionophore A23187-induced cell death in cultured embryonic striatal neurons. Experimental Neurology. 1999;160(2):402–12.

[166] Bachoud-Levi AC, Deglon N, Nguyen JP, Bloch J, Bourdet C, Winkel L, et al: Neuroprotective gene therapy for Huntington's disease using a polymer encapsulated BHK cell line engineered to secrete human CNTF. Human Gene Therapy. 2000;11(12): 1723–9.

[167] Bloch J, Bachoud-Levi AC, Deglon N, Lefaucheur JP, Winkel L, Palfi S, et al: Neuroprotective gene therapy for Huntington's disease, using polymer-encapsulated cells

engineered to secrete human ciliary neurotrophic factor: results of a phase I study. Human Gene Therapy. 2004;15(10):968–75.

[168] Lindvall O, Kokaia Z: Stem cells for the treatment of neurological disorders. Nature. 2006;441(7097):1094–6.

Chapter 6

Antiaging — Effect of Stem Cells on Aging and Stem Cell Aging

Jisook Moon and Sang-Hun Bae

Additional information is available at the end of the chapter

http://dx.doi.org/10.5772/60637

Abstract

Aging is defined broadly as the normal progressive process, consequently leading to growing vulnerability to disease and death. A major challenge lies in dissecting the underlying mechanisms of aging with conventional experiments due to the complexity of and multicontributions to the aging process, reflecting a need for investigation into it in various aspects. For this reason, the age process has currently been subjected to OMICS technologies including genomics, transcriptomics, proteomics, and metabolomics, allowing the exploration of age-related changes in a multifactorial manner. In addition, since age-dependent decline in stem cell function is almost identical to the biological age, stem cells have used to understand "aging" and to investigate key reverse factors for "antiaging". This suggests that a range of new approaches are needed to reveal the unknown biological basis for aging at a variety of different molecular levels using stem cells as a tool of normal aging process and can further apply fundamental aspects in biological aging and longevity.

Keywords: Aging, OMICS, Stem Cells, Transcriptoms, Longevity

1. Introduction

Aging is defined broadly as the normal progressive process, consequently leading to growing vulnerability to disease and death. The fact that the aging process is inevitable yet controllable has made it attractive for the research focusing on age-associated molecular changes. A major challenge lies in dissecting the underlying mechanisms of aging with conventional experi-

INTECH
open science | open minds

© 2015 The Author(s). Licensee InTech. This chapter is distributed under the terms of the Creative Commons Attribution License (http://creativecommons.org/licenses/by/3.0), which permits unrestricted use, distribution, and reproduction in any medium, provided the original work is properly cited.

ments due to the complexity of and multicontributions to the aging process, reflecting a need for investigation into it in various aspects. For this reason, the age process has currently been subjected to OMICS technologies including genomics, transcriptomics, proteomics, and metabolomics, allowing the exploration of age-related changes in a multifactorial manner. This suggests that a range of new approaches are needed to reveal the unknown biological basis for aging at a variety of different molecular levels and can deepen our understanding of fundamental aspects in biological aging and longevity.

The aging process is characterized by gradual, cumulative damages to structure and function of stem cells which exist during the life of organisms. We will discuss here the integrative studies of the stem cell aging and a therapeutic effect of adult stem cells including the umbilical cord blood and the underlying mechanisms of the complex process at diverse molecular levels, with the final goal of practically applying stem cell treatment to the aged for maintaining health over time. In addition, an integrated method, OMICS technology that would help us understand a complex biology of aging will be discussed. Aging can be conceived of as a process that a pool of endogenous stem cells loses progressively its ability to replenish the damaged cells over age. In almost all living organisms, the time-dependent decline in regenerative potential of stem cells is responsible for an increased susceptibility to aging and several age-related diseases. The reduced regenerative capacity of endogenous stem cells has been explained partly by DNA damage, changes in stem cell niches, and activation of tumor suppressor gene. It is unclear; however, to what extent the factors contribute to human ageing, especially stem cell aging, and determine even life span. Its complexity demands new approaches for clarifying the multifactorial processes.

2. OMICS technologies and stem cell aging

Recently, to gain a deep insight into the biology of aging, new high-throughput technologies, also known as "OMICS", are being utilized in a variety of ways to investigate the molecular changes observed during ageing ; OMICS refers to studies with suffix "-omics" designed to collectively characterize and quantify pools of molecules at different levels, including genomics, transcriptomics, proteomics, and metabolomics. A series of experiments using the OMICS have been attempted to establish any link between molecular changes and aging. However, the studies so far on aging heavily rely on blood samples which consist of different cell types and usually focus on one technology of OMICS, thereby placing obstacles in the way of interpretation on the phenomenon or bringing misinterpretation of the complex aging process. Another challenge in the study is tissue-specific changes in gene expression with increasing age, adding more complexity to understanding the process. Accordingly, an alternative approach can be to focus on stem cell aging among diverse hallmarks of the process using OMICS technology. Stem cells serve as endogenous replacements for cells lost to homeostasis and injury through adult life. The regenerative capacity deteriorates in numerous tissues with advancing age, frequently failing to meet the demands of the developing tissues and then leading to multiple ageing-related phenotypes or diseases. As a result, accumulation of damage in the function could be reflected in diverse macromolecules from DNA to metabolite, considered as closest to phenotypes, during aging. The damaged macromolecules in turn disrupt the pathways contributing to stem cell dysfunction during the aging process, resulting in a vicious cycle. In addition, decreased pools of tissue stem cells are likely to be associated with function declines in hematopoiesis, neurogenesis, and myogenesis during aging, suggesting that a key to reverse or delay the aging process resides in deepening our understanding of the adult stem cells. As with other factors for aging, the mechanisms that induce the time-dependent stem cell decline still remain elusive and thereby need to be evaluated in an integrative manner for which OMICS technologies may be appropriate. Understanding the molecular processes involved in stem cell dysfunction may shed light on the causes of aging, eventually employing therapeutic strategies that reverse the decline process. Maintenance of stem cell pools or stem cell rejuvenation holds great therapeutic promise for age-related impairments. For example, heterochronic parabiosis, such as the shared circulatory or physiological system between the young and aged, has been reported to be effective in reversing age-related phenotypes by improving stem cell function. One representative study on the parabiosis showed that the supply of young blood to aged mice ameliorates cognitive impairments by enhancing synaptic plasticity in the brain. Another experiment demonstrated that when exposed to the niche of young mice's muscles, aged mice regenerate impaired satellite cells to restore muscle regenerative potential. Emerging evidence also indicates that reduced regenerative capacity is reversible and that the aging process can be postponed by improving stem cell function to replenish the damaged tissues. This raises the possibility that treatment of stem cells from diverse origins may reengineer the aging-related defects by replacing aberrant stem cells in the aged tissue. With widening applications, metabolomics today is surfacing as a new approach to decipher the regulation of metabolism involved in aging. Metabolites are end products of complex biological events and can be considered as ultimate responses to internal states or external forces, probably providing unrepresented insights into how stem cell declines influence human aging. Oxidative metabolism and the maintenance of mitochondria have been shown to be associated with stem cell aging. Consistently, metabolic states in stem cells play a crucial role in determining whether the cells are bound for proliferation or differentiation; both cell states are mainly associated with mechanisms controlling the balance between glycolysis and oxidative phosphorylation. In addition, clinical studies on aging with metabolic profiles showed the age-specific metabolites having strong correlations, some of which are associated with fatty acid oxidation, underscoring the role of metabolomics in the interpretation of the aging process.

2.1. Transcriptome analysis of neural stem cells during dopamine differentiation

Accordingly, first of all, our study investigated gene expression changes of neural stem cells during differentiation into dopaminergic cells and with increasing passages in a proliferation state, both of which can be seen as aging: differentiation as a part of "chronological aging" and increasing passage as "replicative aging". Neural stem cells showed cell stage-specific patterns of gene expression during differentiation and specific genes participated in neurogenesis by forming a molecular co-expression network. When sustaining a proliferation state, the stem cells induced the expression of genes whose products are involved in phosphorylation, cell proliferation, kinase cascade, response to stress, and signal transduction. As entering into a differentiation stage, the up-regulated genes are mostly related to mitotic cell cycle, mitosis and cell division. At a late differentiation stage, genes for synaptic transmission and regulation of synaptic plasticity were expressed at higher levels. The results clearly showed that as cells age from proliferation to differentiation, different biological processes are involved in stem cell aging, probably generating metabolites unique to cell states.

2.2. Transcriptome analysis of hypoxic effects on placenta-derived cells with increasing passages

Also, we determined the effects of hypoxia or normoxia on the placenta-derived cells with increasing passage based on the transcriptome data. In gene ontology analysis, most genes significantly upregulated under hypoxia were associated with cell proliferation, macromolecule synthesis, metabolic pathway, signaling pathways, and cellular homeostasis, as confirmed by the in vitro result that the hypoxic culture condition enhanced the proliferation capacity. Downregulated genes were enriched for cell death/apoptosis and protein aggregation, supporting the notion that protein homeostasis and balance between proliferation and quiescence are crucial to stem cell aging. These results suggest that under hypoxia the stem cells experience enhanced proliferation and survival, inhibiting cell death and pro-aging pathways. At a late stage, genes that are differentially expressed under hypoxia are enriched for nucleosome assembly and chromatin organization, suggesting the involvement in epigenetic regulation. Lastly, we carried out metabolite profiles in aged mice transplanted with placenta-derived cells. Most of increased metabolites by cell treatment were related to lipid metabolites, which is likely to be associated with unique patterns of gene expression after cell transplantation, encouraging further studies of integrating OMICS data. These findings add weight to the notion that the study of stem cell aging with OMICS is an efficient means for elucidating the biological basis of the aging process. In line with these findings, an effect of human umbilical cord blood infusion, youngest blood we can obtained, on old mice (more than 23 months old) has been investigated in our laboratories, and a human clinical trial using human umbilical cord blood infusion into old subjects is underway. In addition, human placenta-derived MSCs (hpMSCs) have been used as a candidate for antiaging treatment. Our animal studies exhibited better cognitive functions measured 12 weeks after hpMSCs injection. For further translational studies, analyses using OMICS technology is ongoing.

3. Conclusions

Complex physiological changes and individual differences in aging have always challenged the efforts of scientists to understand the normal process, which demands new strategies capable of studying molecular changes in an integrative manner rather than traditional experimental approaches. Through OMICS technologies, it is possible to measure dynamic molecular changes simultaneously at diverse levels with the generation of high-throughput data in different types, consequently facilitating the identification of aging/antiaging biomarkers and thereby preventing age-related diseases. The studies of OMICS have provided novel insights into what molecular pathways determine the progressive and complicated process, although much needs to be clarified. In particular, metabolite profile can propose unprecedented notions of ageing when combined with genomics, transcriptomics, and proteomics. Taken together, the integration and context-dependent interpretation of multidimensional OMICS data is helpful in understanding the complex process of aging.

Author details

Jisook Moon* and Sang-Hun Bae

*Address all correspondence to: jmoon@cha.ac.kr

College of Life Science, Department of Bioengineering, CHA University, Seoul, Korea

References

[1] Signer, R.A. and S.J. Morrison, *Mechanisms that regulate stem cell aging and life span.* Cell Stem Cell, 2013. 12(2): p. 15-–65.

[2] Valdes, A.M., D. Glass, and T.D. Spector, *Omics technologies and the study of human ageing.* Nat Rev Genet, 2013. 14(9): p. 601–7.

[3] Sondheimer, N., et al., *Neutral mitochondrial heteroplasmy and the influence of aging.* Hum Mol Genet, 2011. 20(8): p. 1653–9.

[4] Hannum, G., et al., *Genome-wide methylation profiles reveal quantitative views of human aging rates.* Mol Cell, 2013. 49(2): p. 359–67.

[5] Heyn, H., et al., *Distinct DNA methylomes of newborns and centenarians.* Proc Natl Acad Sci USA, 2012. 109(26): p. 10522–7.

[6] Zahn, J.M., et al., *Transcriptional profiling of aging in human muscle reveals a common aging signature.* PLoS Genet, 2006. 2(7): p. e115.

[7] Morrison, S.J., et al., *The aging of hematopoietic stem cells.* Nat Med, 1996. 2(9): p. 1011–6.

[8] Conboy, I.M., et al., *Rejuvenation of aged progenitor cells by exposure to a young systemic environment.* Nature, 2005. 433(7027): p. 760–4.

[9] Kollman, C., et al., *Donor characteristics as risk factors in recipients after transplantation of bone marrow from unrelated donors: the effect of donor age.* Blood, 2001. 98(7): p. 2043–51.

[10] Enwere, E., et al., *Aging results in reduced epidermal growth factor receptor signaling, diminished olfactory neurogenesis, and deficits in fine olfactory discrimination.* J Neurosci, 2004. 24(38): p. 8354–65.

[11] Cerletti, M., et al., *Regulation and function of skeletal muscle stem cells.* Cold Spring Harb Symp Quant Biol, 2008. 73: p. 317–22.

[12] Villeda, S.A., et al., *Young blood reverses age-related impairments in cognitive function and synaptic plasticity in mice.* Nat Med, 2014. 20(6): p. 659–63.

[13] Sahin, E. and R.A. Depinho, *Linking functional decline of telomeres, mitochondria and stem cells during ageing.* Nature, 2010. 464(7288): p. 520–8.

[14] Chen, H., et al., *Role of SIRT1 and AMPK in mesenchymal stem cells differentiation.* Ageing Res Rev, 2014. 13: p. 55–64.

[15] Yuan, H.F., et al., *SIRT1 is required for long-term growth of human mesenchymal stem cells.* J Mol Med (Berl), 2012. 90(4): p. 389–400.

[16] Kim, H.S., et al., *SIRT3 is a mitochondria-localized tumor suppressor required for maintenance of mitochondrial integrity and metabolism during stress.* Cancer Cell, 2010. 17(1): p. 41–52.

[17] Peserico, A., et al., *A novel AMPK-dependent FoxO3A-SIRT3 intramitochondrial complex sensing glucose levels.* Cell Mol Life Sci, 2013. 70(11): p. 2015–29.

[18] Lawton, K.A., et al., *Analysis of the adult human plasma metabolome.* Pharmacogenomics, 2008. 9(4): p. 383–97.

[19] Yu, Z., et al., *Human serum metabolic profiles are age dependent.* Aging Cell, 2012. 11(6): p. 960–7.

Chapter 7

Mesenchymal Stem Cell Manufacturing for Clinical Use

Franca Fagioli and Ivana Ferrero

Additional information is available at the end of the chapter

http://dx.doi.org/10.5772/61370

Abstract

Mesenchymal stromal/stem cells (MSCs) are adult, self-renewable, multipotent cells that can be isolated in various tissues, most commonly bone marrow (BM) and adipose tissue. Because of their capacity for self-renewal and differentiation into tissues of mesodermal origin and due to their immunomodulatory ability, MSCs are used in many preclinical and clinical studies as possible new therapeutic agents for the treatment of a very broad range of conditions, including heart, hepatic, and neurodegenerative diseases.

Whatever the tissue of origin or the clinical application, the MSCs must be expanded in vitro to obtain sufficient cell numbers, also when multiple doses are needed.

The MSC manufacturing process for clinical use should comply with the principles of Good Manufacturing Practice (GMP). This ensures that cell preparations are produced and controlled, from the collection and manipulation of raw materials, through the processing of intermediate products, to the quality controls, storage, labelling and packaging, and release.

The application of GMP to manufacture medicinal products such as MSCs must ensure that clinical trials are unaffected by inadequate safety, quality, or efficacy arising from unsatisfactory manipulation of the cells.

Keywords: Mesenchymal stem cells, good manufacturing practice, stem cell therapy

1. Introduction

Cell-based therapy has led to the development of new biological medicines to repair, replace, or recover the biological function of damaged tissues and organs [1]. Among cell types used for this purpose, human MSCs are considered as advanced therapy medical products (ATMPs) and should be handled with appropriate controls to ensure their safety, quality, and efficacy

INTECH
open science | open minds

© 2015 The Author(s). Licensee InTech. This chapter is distributed under the terms of the Creative Commons Attribution License (http://creativecommons.org/licenses/by/3.0), which permits unrestricted use, distribution, and reproduction in any medium, provided the original work is properly cited.

as a final drug [2-3]. Mesenchymal stromal or stem cells have become of great interest in these years in the field of regenerative medicine. They were initially referred to as multilineage progenitor cells isolated and culture-expanded from adult human BM that, once in culture, display a heterogeneous morphology and are capable of several subpopulations. They can be expanded ex vivo and readily differentiate into mesodermal cell derivatives [4-5]. Horwitz and the International Society for Cellular Therapy (ISCT) better defined them as "mesenchymal stromal cells" (MSCs) [6].

The translation of research-based protocols into large-scale production of clinical-grade MSCs requires careful analysis to identify and control all critical aspects: source of MSCs and raw materials, culture media, supplements and disposable devices, and quality control tests [7]. The quality and safety of the cell preparations should be ensured by the implementation of a quality system that guarantees the certification and traceability of every batch of material and supply utilized for the procedures, correct utilization and cleaning of instruments, and the locations necessary for stem cell manipulation. Facilities, isolation methods, seeding density, growth factors, and chemicals can all influence the expansion potential and functional properties of MSCs and should be considered throughout the production process. The organization structure, qualification of staff with high levels of expertise, and the appropriate equipment must be implemented in dedicated cleanrooms, currently named "Cell Factories," in compliance with current good manufacturing practice (GMP) standards. The application of GMPs for aseptic production ensures the safety of the final cell therapy product. In 2008, the International Society for Stem Cell Research released a set of recommended guidelines for the development of stem cell–based treatments [8]. These recommendations are focused on the risks involved with stem cell-based therapies, the voluntary informed consent, the patient monitoring and adverse event reporting, and the equality of benefits of stem cell treatments.

2. Body

MSCs are adult, self-renewable, multipotent cells that can be isolated in various tissues, most commonly BM and adipose tissue, but also in fetal tissues. Because of their capacity for self-renewal and differentiation into tissues of mesodermal origin and due to their immunomodulatory ability, MSCs are used in many preclinical and clinical studies as possible new therapeutic agents for the autoimmune or degenerative diseases treatment [9].

The MSC manufacturing process for clinical use should comply with the principles of GMP. This ensures that cell preparations are produced and controlled, from the collection and manipulation of raw materials, through the processing of intermediate products, to the quality controls, storage, labelling and packaging, and release, according to the expected requirements. During the whole production process, critical steps should be known and described. A thorough risk analysis during all phases of production and control ensures a final product with the expected quality, both for patients and for the regulatory authorities responsible for the controls.

2.1. MSC source

BM is still the most common source of MSCs. However, during the last two decades alternative and more accessible tissue sources of MSCs have been identified.

MSCs have been isolated from fat, deciduous teeth, placenta, and umbilical cord blood, showing comparable features to BM-derived cells [10]. In particular, fresh umbilical cord blood is the third common source for isolating MSCs for clinical use, but the success rate in isolating and further expanding MSCs depends on the volume of blood collected, the cell content, and the time between collection and processing [12]. Related cell sources, including neonatal tissues as amniotic fluid [11], placenta, and Wharton's jelly (WJ), are interesting for their ability to maintain their self-renewal capacity *in vitro* for a long period, showing a higher pluripotency capacity as compared to BM-MSCs [13-15].

Adipose-derived stem cells (ASCs), if compared to BM MSCs, appear to have higher frequencies [16], higher proliferative capacity [16], and to be more stable in culture with a lower senescence ratio [18]. The methods to isolate MSCs from lipoaspirates are standardized, even with automated device to separate the tissue fraction of fat from the contaminating fluids [19]. ASCs have been tested in preclinical studies for the treatment of a variety of clinical conditions, including Crohn's disease, muscular dystrophy, myocardial infarction, spinal cord injury, diabetes mellitustype1, breast reconstruction, and facial lipoatrophy [20].

2.2. MSC isolation and expansion

Several different methods for isolating MSCs have been described. After harvesting and, if necessary as in case of fat, dissociation of tissue, MSCs could be seeded in flask with or without gradient separation (Ficoll or Percoll), or after immunomagnetic enrichment [21]. The paper of Mareschi showed that direct selection of MSCs from BM cells by adhesion to culture plastic is a more advantageous method compared to MSCs obtained by gradient separation [22], resulting also in a significantly higher level of HLA-DR [23]. Cell-plating density is a crucial issue for MSC expansion. The use of very low density ensures a high proliferative potential [24]. However, plating densities of 10 and 100 cells/cm^2 did not produce any expansion [22]. Limiting the number of cell population doubling (CPD) to less than 20 avoids the progressive senescence of expanded MSCs [23].

The use of reagents and supplements clinical grade and suitable for clinical application must be considered for the *in vitro* expansion. Only few clinical grade reagents are commercially available. Animal serum–free media, human serum, and other supplements of human origin have been investigated. Human platelet lysate (HPL) has been demonstrated to be safe and efficient as MSC culture supplement for robust MSC expansion and is now considered a good fetal bovine serum (FBS) substitute [25-26]. In particular, Pathogen Inactivated HPL represents a good GMP-compliant alternative to FBS [27].

2.3. MSCs as ATP

The European Regulation 1394/2007 introduced a new classification for cell therapy products as medicinal products: the tissue-engineered products [28]. The Regulation defines Advanced

Therapy Medicinal Products (ATMPs) as the gene therapy medicinal products, the somatic cell therapy medicinal products, and the tissue-engineered products. These products have specific pharmacologic, metabolic, and immunologic activities and the potential for treating a variety of disorders. For these reasons, cellular products for ATMPs, prepared on a routine basis, must meet the same stringent conditions required for drugs preparation and commercialization. In particular, they must be manipulated according to the GMP [29], and they require preclinical testing in Good Laboratory Practice (GLP) [30-31] and clinical trials conducted in Good Clinical Practice (GCP) before being commercialized [32].

MSCs represent a promising strategy for the development of new therapeutic strategies due to the numerous applications proposed for their use [33]. Thanks to their intrinsic properties, MSCs represent an attractive candidate for clinical applications [34-36]. As ATMPs, MSCs also must satisfy all the above-mentioned requirements. The expansion protocol of MSCs for clinical use should comply with the principles of GMP and take in account all critical steps of the process. The quality and safety of the expanded MSCs must be maintained throughout their production and quality control cycle, ensuring their safe use in the patient.The main critical steps are the collection and manipulation of raw materials; the processing of intermediate products; the quality control tests; storage, labelling, and packaging; and release. The validation of the process and a thorough risk analysis during all phases of production ensures a final product with the expected quality, both for patients and for the regulatory authorities responsible for controls [37].

2.4. GMP production of MSCs

MSCs represent a small fraction (0.001-0.01%) of the BM cell population; therefore, to obtain a sufficient number of cells, they must be extensively *ex vivo* expanded.

One of the most significant properties that make MSCs a special device for cell-based therapeutic approaches is their capability to expand in culture for several passages, without compromising their pluripotent potential. MSCs can easily be expanded in culture thanks to their capacity to adhere and proliferate maintaining their immunophenotypic characteristics and functions as multipotent cells.

Several papers discussed in the last years the standardization of MSC culture conditions for clinical use. As reported in the review by Ikebe, a total of 47 reports, describing MSC-isolation methods, were found in literature published from January 2007 to 2013 [38].

The recent paper of Wuchter summarizes a consensus meeting between researchers, clinicians, and regulatory experts on standard quality requirements for MSC production [39]. The manufacture of sterile products, such as the ATMPs, is subject to special requirements in order to minimize the risks of microbiological contamination. Maintaining the cells in culture for several passages means subjecting them to the risk of contamination. The production process of MSCs must be performed through a GMP protocol based on standard operating procedures (SOPs) and documented according to GMP [40-41].

As mentioned in Annex 1 of GMP [42], the manufacture of sterile products should be carried out in clean areas, laboratory named cleanrooms or cell factories, in which the concentration

http://dx.doi.org/10.5772/61370

of airborne particulates is maintained and controlled. The key elements for designing a cleanroom according to GMP requirements are the presence of HEPA filters, a room air change rate of >20 per hour, the temperature and humidity control, and the air pressure differential. The appropriate cleanliness standard can be maintained also through an appropriate management of personnel, equipment, and materials. Premises and equipment should be designed in order to minimize the accumulation of particles or microorganisms and to permit the appropriate cleaning.

For the manufacture of sterile medicinal products, four grades can be distinguished:

- Grade C and D: clean areas for carrying out manufacturing steps not so critical.
- Grade B, for aseptic manipulation, is the background environment for grade A.
- Grade A, the local zone for high-risk operations, where the product is manipulated in an open system. This condition is normally the laminar airflow biosafety cabinet.

Annex 1 also gives the recommended limits for microbiological monitoring of clean areas during aseptic operations.

The concept of quality assurance and quality management are related to GMP requirements.

As reported in the Chapter 1 of GMP guidelines [43], the quality management covers all matters, which individually or collectively influence the quality of a product.

A GMP-compliant quality system for the manufacture of medicinal products should ensure that cell preparations are controlled, from the collection and manipulation of raw materials, through the processing of intermediate products, to the quality controls, storage, labelling and packaging, and release. During the whole production process, critical steps should be known and described.

An appropriate quality system should cover all the following aspects:

- Roles and responsibilities. All the staff must be qualified and adequate. A qualified person must certify the release of the medicinal products. Competencies, training and retraining must be scheduled.
- Premises and equipments.
- Documentation.
- Raw materials, reagents, and suppliers.
- Quality control on intermediate products, in-process controls, and validations.
- Activities in outsourcing.
- Deviations, corrective/preventive actions, and change control.
- Storage, distribution, and labelling.
- Self-inspection and/or quality audits.

Good documentation constitutes an essential part of the quality assurance system. Documents may exist as paper-based or electronic format. The main documents, as required by GMP are as follows:

- Site Master File, describing the Facility and the GMP activities.
- SOPs, detailed description of various processes.
- Working instructions, giving instructions on technical methods or activities.
- Records, registration, and reports, providing evidence of actions and steps.
- Certificate of analysis, providing a summary of quality control tests with the evaluation of compliance of the product batch.

Approved procedures and instructions must be written in a clear and unambiguous form, and the records must demonstrate that all the steps required by the defined procedures and instructions were carried out by qualified operators.

2.5. Quality control

Quality Control is that part of GMP which is concerned with sampling, specifications, and testing [44], also including outsourcing activities. The quality control process should be validated to confirm that the analytical procedure employed for a specific test is suitable for its intended use.

The recent paper of Torre and colleagues provides the minimal quality requirements for the MSC production and its delivery for clinical use, to guarantee the safety of the final cell therapy product. For this purpose, the document evaluates the most important steps of GMP-compliant MSC production: the isolation and expansion process; the validation phase of the process, including all quality controls for the characterization, functionality, potency, and safety of MSCs; and the quality control at the batch release to guarantee the safety of patient infusion [37].

An interesting approach to the GMP is validation. The production process should follow a validation protocol, also considering instruments, supplies and reagents, and defining roles and responsibilities of each step [45]. A validation protocol on a minimum of three expansion procedures should be implemented to standardize the best culture condition and to demonstrate the safety and feasibility of the production protocol. It is a requirement of GMP that manufacturers identify what validation work is needed to prove control of the critical aspects of their particular operations. Significant changes to the facilities, the equipment, and the processes, which may affect the quality of the product, should be validated. A risk assessment approach should be used to determine the scope and extent of validation.

According to International Conference on HarmonizationQ2 (ICH Q2) Guidelines [45] and the European (EU) Pharmacopoeia [47], the quality control (QC) process should be validated to confirm that the analytical procedure employed for a specific test is suitable for its intended use.

Quality control analytical methods could be validated demonstrating the accuracy, the specificity, the limit of detection, and the precision [45, 48-49].

In the validation phase of the process, additional assays to test immunomodulation, senescence, and genomic stability, clonogenicity should be carried out. Although tumour formation during the validation process has not been reported in ongoing clinical trials using MSCs [50], tumorigenesis and karyotype [23, 51] should be also tested during validation.

2.6. Clinical trials

Today, numerous MSC preparations from academic institutions and private companies are being investigated. The review paper of Wang summarized the currently completed clinical trials, registered with clinicaltrials.gov, using MSC to treat various diseases. The first clinical trial using culture-expanded MSCs was carried out in 1995 and 15 patients became the recipients of the autologous cells [52]. Since then, a great number of clinical trials have been conducted to test the feasibility and efficacy of MSC therapy.

Today more than 450 clinical trials testing MSC are currently published on the international registry www.clinicaltrials.gov, both recruiting or completed. Most trials are Phase I (safety studies) or Phase I/II studies. Only a small number of these trials are in Phase III.

Expanded MSCs have been used to treat several diseases, from bone/cartilage disorder, cardiovascular diseases, autoimmune diseases, and neurodegenerative disorders [12].

The experience of our group has been focalized to research a new possible strategy for amyotrophic lateral sclerosis (ALS) aimed at cell replacement and neuroprotection.

The long term study of showed the safety of MSC transplantation in the central nervous system during a follow-up of nearly 9 years, and is in support of applying MSC-based cellular clinical trials to neurodegenerative disorders [53-55].

The application of GMP to the manufacturing of investigational medicinal products for clinical trials must ensure that trial subjects are not placed at risk, and that the results of clinical trials are unaffected by inadequate safety, quality, or efficacy arising from unsatisfactory manipulation of the cells. Annex 13 of GMP guidelines is the reference guide for this step [56].

3. Conclusion

Stem cell therapy has evolved over the past years, thanks to the great efforts of researchers to define and implement new methods for MSC expansion, according to ATMP requirements.

The standardization of protocols and procedures, both for production and quality control, within the collaboration of different groups, will lead to the definition of standardized protocols to be considered during the process of expansion and release of MSC as ATMPs for the application of clinical trials with MSCs.

Author details

Franca Fagioli[1*] and Ivana Ferrero[1,2]

*Address all correspondence to: franca.fagioli@unito.it

1 Paediatric Onco-Hematology, Stem Cell Transplantation and Cellular Therapy Division, City of Health and Science of Turin., Italy

2 Department of Public Health and Pediatrics, University of Turin, Italy

References

[1] Gálvez P, Clares B, Hmadcha A, Ruiz A, Soria B. Development of a cell-based medicinal product: regulatory structures in the European Union. Br Med Bull. 2013;105:85-105. doi:10.1093/bmb/lds036

[2] Commission Directive 2009/120/EC of 14 September 2009 amending Directive 2001/83/EC of the European Parliament and of the Council on the Community code relating to medicinal products for human use as regards advanced therapy medicinal products. Official J. Eur. Union L 242.

[3] Sensebé L. Clinical grade production of mesenchymal stem cells. Biomed Mater Eng. 2008;18:S3-10. Review. PMID:18334718.

[4] Pittenger MF, Mackay AM, Beck SC, Jaiswal RK, Douglas R, Mosca JD, Moorman MA, Simonetti DW, Craig S, Marshak DR. Multilineage potential of adult human mesenchymal stem cells. Science. 1999;284:143-147. PMID:10102814.

[5] Horwitz EM, Gordon PL, Koo WK, Marx JC, Neel MD, McNall RY, Muul L, Hofmann T. Isolated allogeneic bone marrow-derived mesenchymal cells engraft and stimulate growth in children with osteogenesisimperfecta: implications for cell therapy of bone. ProcNatlAcadSci USA. 2002;99:8932e7. PMID:12084934.

[6] Horwitz EM, Le Blanc K, Dominici M, Mueller I, Slaper-Cortenbach I, Marini FC, Deans RJ, Krause DS, Keating A. International society for cellular therapy. Clarification of the nomenclature for MSC: The international society for cellular therapy position statement. Cytotherapy 2005;7:393-395.PMID:16236628.

[7] Sensebé L, Bourin P. Mesenchymal stem cells - The challenge of a good therapeutic product. Med Sci (Paris). 2011;27:297-302. doi: 10.1051/medsci/2011273297.

[8] Hyun I, Lindvall O, Ahrlund-Richter L, Cattaneo E, Cavazzana-Calvo M, Cossu G, De Luca M, Fox IJ, Gerstle C, Goldstein RA, Hermerén G, High KA, Kim HO, Lee HP, Levy-Lahad E, Li L, Lo B, Marshak DR, McNab A, Munsie M, Nakauchi H, Rao M, Rooke HM, Valles CS, Srivastava A, Sugarman J, Taylor PL, Veiga A, Wong AL,

Zoloth L, Daley GQ. New ISSCR guidelines underscore major principles for responsible translational stem cell research. Cell Stem Cell. 2008;3:607-609. doi:10.1016/j.stem.2008.11.009. Review.

[9] Abdi R, Fiorina P, Adra CN, Atkinson M, Sayegh MH. Immunomodulation by mesenchymal stem cells: a potential therapeutic strategy for type 1 diabetes. Diabetes. 2008 Jul;57(7):1759-67.

[10] Parolini O, Alviano F, Bagnara GP, Bilic G, Bühring HJ, Evangelista M, Hennerbichler S, Liu B, Magatti M, Mao N, Miki T, Marongiu F, Nakajima H, Nikaido T, Portmann-Lanz CB, Sankar V, Soncini M, Stadler G, Surbek D, Takahashi TA, Redl H, Sakuragawa N, Wolbank S, Zeisberger S, Zisch A, Strom SC. Concise review: isolation and characterization of cells from human term placenta: outcome of the first international Workshop on Placenta Derived Stem Cells. Stem Cells. 2008;26:300-311. Review. PMID:17975221.

[11] De Coppi P, Bartsch G Jr, Siddiqui MM, Xu T, Santos CC, Perin L, Mostoslavsky G, Serre AC, Snyder EY, Yoo JJ, Furth ME, Soker S, Atala A. Isolation of amniotic stem cell lines with potential for therapy. Nat Biotechnol. 2007;25:100-106. PMID:17206138.

[12] Sharma RR, Pollock K, Hubel A, McKenna D. Mesenchymal stem or stromal cells: a review of clinical applications and manufacturing practices. Transfusion. 2014;54:1418-1437. doi:10.1111/trf.12421.

[13] Wang S, Qu X, Zhao RC. Clinical applications of mesenchymal stem cells. J HematolOncol. 2012;5:19. doi:10.1186/1756-8722-5-19.

[14] Troyer DL, Weiss ML. Concise review: Wharton's jelly-derived cells are a primitive stromal cell population. Stem Cells. 2008;26: 591-599. PMID:18065397.

[15] Caruso M, Evangelista M, Parolini O. Human term placental cells: phenotype, properties and new avenues in regenerative medicine. Int J Mol Cell Med. 2012;1:64-74. PMID:24551761.

[16] Moseley TA, Zhu M, Hedrick MH. Adipose-derived stem and progenitor cells as fillers in plastic and reconstructive surgery. PlastReconstr Surg. 2006;118(3 Suppl): 121S-128S. PMID:16936551.

[17] Izadpanah R, Trygg C, Patel B, Kriedt C, Dufour J, Gimble JM, Bunnell BA. Biologic properties of mesenchymal stem cells derived from bone marrow and adipose tissue. J Cell Biochem. 2006;99:1285-1297. PMID: 16795045.

[18] Kern S, Eichler H, Stoeve J, Klüter H, Bieback K. Comparative analysis of mesenchymal stem cells from bone marrow, umbilical cord blood, or adipose tissue. Stem Cells. 2006;24:1294-1301.PMID:16410387.

[19] Katz AJ, Hedrick MH, Llull R, Futrell JW. A novel device for the simple and efficient refinement of liposuctioned tissue. PlastReconstr Surg. 2001;107:595-597.

[20] Sousa BR, Parreira RC, Fonseca EA, Amaya MJ, Tonelli FM, Lacerda SM, Lalwani P, Santos AK, Gomes KN, Ulrich H, Kihara AH, Resende RR. Human adult stem cells from diverse origins: an overview from multiparametricimmunophenotyping to clinical applications. Cytometry 2014;85:43-77.PMID:24700575.

[21] Sensebé L, Gadelorge M, Fleury-Cappellesso S. Production of mesenchymal stromal/ stem cells according to good manufacturing practices: a review. Stem Cell Res Ther. 2013;4:66.doi:10.1186/scrt217.

[22] Mareschi K, Rustichelli D, Calabrese R, Gunetti M, Sanavio F, Castiglia S, Risso A, Ferrero I, Tarella C, Fagioli F. Multipotent mesenchymal stromal stem cell expansion by plating whole bone marrow at a low cellular density: a more advantageous method for clinical use. Stem Cells International 2012:920581. doi:10.1155/2012/920581.

[23] Tarte K, Gaillard J, Lataillade JJ, Fouillard L, Becker M, Mossafa H, Tchirkov A, Rouard H, Henry C, Splingard M, Dulong J, Monnier D, Gourmelon P, Gorin NC, Sensebé L. SociétéFrançaise de Greffe de Moelle et ThérapieCellulaire. Clinical-grade production of human mesenchymal stromal cells: occurrence of aneuploidy without transformation. Blood. 2010;115:1549-1553. doi:10.1182/blood-2009-05-219907.

[24] Colter DC, Sekiya I, Prockop DJ. Identification of a subpopulation of rapidly self-renewing and multipotential adult stem cells in colonies of human marrow stromal cells. ProcNatlAcadSci U S A. 2001;98:7841-7845. PMID:11427725.

[25] Schallmoser K, Bartmann C, Rohde E, Reinisch A, Kashofer K, Stadelmeyer E, Drexler C, Lanzer G, Linkesch W, Strunk D. Human platelet lysate can replace fetal bovine serum for clinical-scale expansion of functional mesenchymal stromal cells. Transfusion. 2007;47:1436-1446. PMID:17655588.

[26] TrojahnKølle SF, Oliveri RS, Glovinski PV, Kirchhoff M, Mathiasen AB, Elberg JJ, Andersen PS, Drzewiecki KT, Fischer-Nielsen A. Pooled human platelet lysate versus fetal bovine serum-investigating the proliferation rate, chromosome stability and angiogenic potential of human adipose tissue-derived stem cells intended for clinical use. Cytotherapy. 2013;15:1086-1097. doi:10.1016/j.jcyt.2013.01.217.

[27] Castiglia S, Mareschi K, Labanca L, Lucania G, Leone M, Sanavio F, Castello L, Rustichelli D, Signorino E, Gunetti M, Bergallo M, Bordiga AM, Ferrero I, Fagioli F. Inactivated human platelet lysate with psoralen: a new perspective for mesenchymal stromal cell production in Good Manufacturing Practice conditions. Cytotherapy. 2014;16:750-763. doi:10.1016/j.jcyt.2013.12.008.

[28] Regulation (EC) N. 1394/2007 of the European Parliament and of the Council of 13 November 2007 on advanced therapy medicinal products and amending Directive 2001/83/EC and Regulation (EC) N. 726/2004.

[29] EudraLex. Volume 4: Good manufacturing practice (GMP) Guidelines for medicinal products for human and veterinary use, laid down in Commission Directives 91/356/ EEC, as amended by Directive 2003/94/EC, and 91/412/EEC.

[30] Directive 2004/9/EC of the European Parliament and of the Council of 11 February 2004 on the inspection and verification of good laboratory practice (GLP). Official Journal of the European Union. L 50/28. 20/02/2004.

[31] Directive 2004/10/EC of the European Parliament and of the Council of 11 February 2004 on the harmonisation of laws, regulations and administrative provisions relating to the application of the principles of good laboratory practice and the verification of their applications for tests on chemical substances (codified version). Official Journal of the European Union. L 50/44. 20/02/2004.

[32] European Commission. Detailed guidance on the request to the competent authorities for authorisation of a clinical trial on a medicinal product for human use, the notification of substantial amendments and the declaration of the end of the trial (CT-1). Official Journal of the European Union 2010/C 82/01.

[33] De Girolamo L, Lucarelli E, Alessandri G, Avanzini MA, Bernardo ME, Biagi E, Brini AT, D'Amico G, Fagioli F, Ferrero I, Locatelli F, Maccario R, Marazzi M, Parolini O, Pessina A, Torre ML. Italian Mesenchymal Stem Cell Group. Mesenchymal stem/stromal cells: a new cells as drugs' paradigm. Efficacy and critical aspects in cell therapy. Curr Pharm Des. 2013;19:2459-2473. PMID:23278600.

[34] Patel DM, Shah J, Srivastava AS. Therapeutic potential of mesenchymal stem cells in regenerative medicine. Stem Cells Int 2013:496218. doi:10.1155/2013/496218.

[35] Motaln H, Schichor C, Lah TT. Human mesenchymal stem cells and their use in cell-based therapies. Cancer. 2010;116:2519-2530. doi:10.1002/cncr.25056.

[36] Yi T, Song SU. Immunomodulatory properties of mesenchymal stem cells and their therapeutic applications. Arch Pharm Res. 2012,35:213-221. doi:10.1007/s12272-012-0202-z.

[37] Torre ML, Lucarelli E, Guidi S, Ferrari M, Alessandri G, de Girolamo L, Pessina A, Ferrero I. Ex-vivo expanded mesenchymal stromal cell minimal quality requirements for clinical application. Stem Cells Dev. 2015 Mar 15;24(6):677-85. PMID:25517941.

[38] Ikebe C, Suzuki K. Mesenchymal stem cells for regenerative therapy: optimization of cell preparation protocols. BioMed Res Int. 2014:951512. doi:10.1155/2014/951512.

[39] Wuchter P, Bieback K, Schrezenmeier H, Bornhäuser M, Müller LP, Bönig H, Wagner W, Meisel R, Pavel P, Tonn T, Lang P, Müller I, Renner M, Malcherek G, Saffrich R, Buss EC, Horn P, Rojewski M, Schmitt A, Ho AD, Sanzenbacher R, Schmitt M. Standardization of Good Manufacturing Practice-compliant production of bone marrow-derived human mesenchymal stromal cells for immunotherapeutic applications. Cytotherapy. 2015;17:128-139. doi:10.1016/j.jcyt.2014.04.002.

[40] European Medicine Agency. Guideline on Human Cell-based Medicinal Products. EMEA/CHMP/410869/2006. London. 2007.

[41] EudraLex. Volume 4: Good manufacturing practice (GMP) Guidelines for medicinal products for human and veterinary use, laid down in Commission Directives 91/356/

EEC, as amended by Directive 2003/94/EC, and 91/412/EEC. Annex 2. Manufacture of Biological active substances and Medicinal Products for Human Use. 2013. Available from http://ec.europa.eu/health/documents/eudralex/vol-4/index_en.htm

[42] EudraLex. Volume 4: Good manufacturing practice (GMP) Guidelines for medicinal products for human and veterinary use, laid down in Commission Directives 91/356/EEC, as amended by Directive 2003/94/EC, and 91/412/EEC. Annex 1 Manufacture of Sterile Medicinal Products, rev. 2008. Available from http://ec.europa.eu/health/documents/eudralex/vol-4/index_en.htm

[43] EudraLex. Volume 4: Good manufacturing practice (GMP) Guidelines for medicinal products for human and veterinary use, laid down in Commission Directives 91/356/EEC, as amended by Directive 2003/94/EC, and 91/412/EEC. Chapter 1 Pharmaceutical Quality System. Rev. 2013. Available from http://ec.europa.eu/health/documents/eudralex/vol-4/index_en.htm.

[44] EudraLex. Volume 4: Good manufacturing practice (GMP) Guidelines for medicinal products for human and veterinary use, laid down in Commission Directives 91/356/EEC, as amended by Directive 2003/94/EC, and 91/412/EEC.Chapter 6: Quality Control. Rev. 2014. Available from http://ec.europa.eu/health/documents/eudralex/vol-4/index_en.htm.

[45] Rustichelli D, Castiglia S, Gunetti M, Mareschi K, Signorino E, Muraro M, Castello L, Sanavio F, Leone M, Ferrero I, Fagioli F. Validation of analytical methods in compliance with good manufacturing practice: a practical approach. J Transl Med. 2013;11:197. doi:10.1186/1479-5876-11-197.

[46] International Conference on Harmonization (ICH). Harmonised Tripartite Guideline. Validation of analytical procedures: text and methodology Q2 (R1). 2005.

[47] European Pharmacopoeia.8th edition. Strasbourg, France. European Directorate for the Quality of Medicines & HealthCare (EDQM). 2013. Available from https://www.edqm.eu/en/european-pharmacopoeia-8th-edition-1563.html.

[48] Gunetti M, Castiglia S, Rustichelli D, Mareschi K, Sanavio F, Muraro M, Signorino E, Castello L, Ferrero I, Fagioli F. Validation of analytical methods in GMP: the disposable Fast Read 102® device, an alternative practical approach for cell counting. J Transl Med. 2012;10:112. doi:10.1186/1479-5876-10-112.

[49] Gálvez P, Clares B, Bermejo M, Hmadcha A, Soria B. Standard requirement of a microbiological quality control program for the manufacture of human mesenchymal stem cells for clinical use. Stem Cells Dev. 2014;23:1074-1083.doi:10.1089/scd.2013.0625.

[50] Bernardo ME, Zaffaroni N, Novara F, Cometa AM, Avanzini MA, Moretta A, Montagna D, Maccario R, Villa R, Daidone MG, Zuffardi O, Locatelli F. Human bone marrow derived mesenchymal stem cells do not undergo transformation after long-term

in vitro culture and do not exhibit telomere maintenance mechanisms. Cancer Res. 2007;67:9142-9149.PMID:17909019.

[51] Zhang ZX, Guan LX, Zhang K, Wang S, Cao PC, Wang YH, Wang Z, Dai LJ. Cytogenetic analysis of human bone marrow-derived mesenchymal stem cells passaged in vitro. Cell Biol Int. 2007;31:645-648. PMID:17207646.

[52] Lazarus HM, Haynesworth SE, Gerson SL. et al. Ex vivo expansion and subsequent infusion of human bone marrow-derived stromal progenitor cells: implications for therapeutic use. Bone Marrow Transplant 1995;16: 557-564.PMID:8528172.

[53] Mazzini L, Mareschi K, Ferrero I, Miglioretti M, Stecco A, Servo S, Carriero A, Monaco F, Fagioli F. Mesenchymal stromal cell transplantation in amyotrophic lateral sclerosis: a long-term safety study. Cytotherapy. 2012;14:56-60. doi: 10.3109/14653249.2011.613929.

[54] Mazzini L, Ferrero I, Luparello V, Rustichelli D, Gunetti M, Mareschi K, Testa L, Stecco A, Tarletti R, Miglioretti M, Fava E, Nasuelli N, Cisari C, Massara M, Vercelli R, Oggioni GD, Carriero A, Cantello R, Monaco F, Fagioli F. Mesenchymal stem cell transplantation in amyotrophic lateral sclerosis: a phase I clinical trial. Exp Neurol. 2010;223:229-237. doi:10.1016/j.expneurol.2009.08.007.

[55] Mazzini L, Vercelli A, Ferrero I, Mareschi K, Boido M, Servo S, Oggioni GD, Testa L, Monaco F, Fagioli F. Stem cells in amyotrophic lateral sclerosis: state of the art. Expert Opin Biol Ther. 2009;9:1245-1258. doi:10.1517/14712590903186956.

[56] EudraLex. Volume 4: Good manufacturing practice (GMP) Guidelines for medicinal products for human and veterinary use, laid down in Commission Directives 91/356/EEC, as amended by Directive 2003/94/EC, and 91/412/EEC. Annex 13 Investigational Medicinal Products. Rev. 2010. Avaliable from http://ec.europa.eu/health/documents/eudralex/vol-4/index_en.htm.

Chapter 8

Trends in Mesenchymal Stem Cells' Applications for Skeletal Muscle Repair and Regeneration

A.R. Caseiro, T. Pereira, P.J. Bártolo, J.D. Santos,
A.L. Luís and A. C. Maurício

Additional information is available at the end of the chapter

http://dx.doi.org/10.5772/60919

Abstract

Skeletal muscle injuries are quite frequent in traumatic scenarios, such as war injuries or road- or work-related accidents. The skeletal muscle has good regenerative ability, but the extent or recurrence of muscle injury might impair complete structural and functional recovery. Severe tissue loss overwhelms skeletal muscle's intrinsic regenerative capabilities and culminates in the development of noncontractile fibrous tissue scar. Conservative RICE -based and surgical treatments show limited efficacy in terms of improving these severe cases outcomes, pressing the need for new approaches on skeletal muscle's therapy. Since the first suggestions of the potential of mesenchymal stem cells for regenerative medicine and tissue engineering, many applications have been explored for a variety of tissues and diseases, including the skeletal muscle, which is the focus of this literature review.

Current research has focused on the influence of nonmuscular MSCs on promoting tissue healing and limiting fibrotic scar formation, as well as on the modulation of the inflammatory response to injury. The most popular source of MSCs is, without a doubt, the bone marrow. However, MSC populations are present in virtually all body tissues, and alternative sources have been proposed, such as the adipose tissue, synovial membrane, dental pulp, and even umbilical cord tissue.

MSCs from various sources have been demonstrated as capable of *in vitro* differentiation into myogenic lineages, through adequate stimuli, displaying phenotypical markers of native skeletal muscle cells. In addition, *in vivo* applications suggest they are capable of integrating host muscular tissues, even when delivered systemically.

INTECH
open science | open minds

© 2015 The Author(s). Licensee InTech. This chapter is distributed under the terms of the Creative Commons Attribution License (http://creativecommons.org/licenses/by/3.0), which permits unrestricted use, distribution, and reproduction in any medium, provided the original work is properly cited.

MSCs are capable of secreting a wide range of active molecules to their surrounding media, including growth factors, chemokines, and cytokines. Most of these growth factors have been associated to the skeletal muscle's regenerative process, and their efficiency has been demonstrated to increase when applied in spatial and temporal coordination. Hence, the combination of molecules secreted by MSCs gained interest as modulator of inflammatory, fibrotic, and regenerating events. This has been proposed as possibly the primary mode of action of undifferentiated MSCs into a lesion site, providing controlled release of such components. Concurrently to the implantation of undifferentiated cells, it has also been hypothesized that the application of secretion products alone (termed as Conditioned Medium) display similar if not improved effects on skeletal tissue regeneration, as it does in other damaged tissues.

Keywords: Skeletal muscle, skeletal muscle regeneration, cellular therapies, mesenchymal stem cells, biomaterials, secretome

1. Introduction

Skeletal muscle accounts for nearly half of the human body mass [1], and inherited and acquired pathologies are often observed in clinical practice.

Given their impact on quality and life expectancy of patients, severe forms of degenerative muscular diseases, such as Duchenne muscular dystrophy (DMD), have been one of the hot topics of skeletal muscle regeneration research, and encouraging results have been obtained through the application of mesenchymal stem cells (MSCs), giving hope for the development of new therapies that can effectively improve the quality of life of affected patients [2–7].

Acquired muscle affections are seemingly more common in active humans, greatly associated to sports practice, but also quite frequent in other traumatic sceneries, such as road or work-related accidents or war injuries [8–11]. Muscle damage can result from ischemia and denervation, to contusion, sprain damage, laceration, avulsion, and other severe tissue losses.

Skeletal muscle has a good regenerative ability, but the extent or recurrence of these insults might impair complete myofibers regeneration, limiting structural and functional recovery of the affected muscle groups. Severe tissue loss usually supplants skeletal muscle's intrinsic regenerative capabilities [8] and culminates in the development of noncontractile fibrous tissue scar [12]. Other well-known factor to impact the intrinsic capacity of skeletal muscle to respond to injury events is the age of the patient [13], affecting both intrinsic cellular mechanisms and their involving niche, hindering their effectiveness upon activation [14]. The regeneration potential of skeletal muscle depends on a multitude of cell types that, upon exposure to specific cues, cooperate to regenerate the damaged tissue, generating a coordinated tissue response [15]. Under particular conditions, such as chronic diseases and aging, the ability of these cells to support the regenerative response declines, leading to maladaptive responses, e.g., the formation of fibrotic scars and fatty infiltration [15].

http://dx.doi.org/10.5772/60919

Current recommendations for skeletal muscle lesions management rely on empirical application of conservative RICE-based (rest, ice/cold, compression and elevation) and surgical treatments [10, 16] but show limited efficacy in terms of improving severe cases outcomes, pressing the need for new approaches on skeletal muscle´s therapy.

Presently, biomedical research is working in various fronts toward complete restoration of structure and function of damaged muscles, converging efforts in the areas of biomaterial development, cell systems applications, and bioactive molecules aiming at filling the defect and recovering the esthetics of the body part, as well as its function.

One of the strategies being intensely explored involves the application of muscle resident and nonmuscular stem cells in search for faster and more effective recovery from severe injuries, restoring both tissue structure and function [17, 18].

2. Skeletal muscle structure and intrinsic healing mechanisms

The basis of skeletal muscle structure and regeneration have been extensively revised in literature, and only a brief description and emphasis to strategic "key points" will be given herein [9, 16].

Skeletal muscle is composed of a mixture of muscle-specific cells, nerves, blood vessels, and connective tissue support matrix. Skeletal muscle tissue-specific cells are multinucleated structures holding complex and highly organized contraction machinery enclosed within the plasma membrane (sarcolemma), and a single cell is termed as myofiber. According to their contractile properties, myofibers can be classified into three types. Type 1 myofibers are slow contracting and fatigue resistant, type 2A myofibers are fast contracting and have intermediate resistance to fatigue, and type 2B myofibers are fast contracting and have poor fatigue resistance. The function and training of a specific muscle or muscle group determine their composition in terms of fiber type content.

The extracellular matrix (ECM) supporting the myofibers (basal lamina or basement membrane) is composed of a vast set of proteins, such collagen, fibronectin, laminin, and other glycoproteins [9]. Myofibers are bound together by connective tissue sheaths (the endomysium, perimysium, and epimysium) associating them at three upscaling levels from involving a single myofiber to the whole muscle belly [16].

The healing process following skeletal muscle injury is classically divided into three interrelated and time-dependent phases, conveying the destruction, repair, and remodeling of the affected tissue. The first phase (destruction phase) is defined by the rupture and necrosis and degeneration of the myofibers (mainly mediated by alterations of the sarcolemma and loss of calcium homeostasis) and associated neurovascular structures and ECM, by the formation of a hematoma (between the damaged/ruptured and retracted muscle cells) and the initiation of the inflammatory cells response [9, 16]. Other authors distinguish a primary hemostatic stage, preceding the inflammatory reaction [19]. The inflammatory phase becomes evident within 24 hours after injury and comes up until approximately 3 days after the event. It is defined by a sequential influx of neutrophils and macrophages to the site, engaging in the phagocytosis of the debris on site and the release of a cascade of mediatory cytokines [9, 20].

The following phase (repair phase) includes the phagocytosis of the debris resulting from the damaged/necrotized tissue, the kickoff of the regeneration of the myofibers, and the production of a connective-tissue scar by migrating fibroblasts and neurovascular regrowth [16].

At this stage, satellite cells (SCs) assume a preponderant role. These cells, firstly identified in the early 1960s [21], constitute a population of myogenically committed but undifferentiated cells, residing between the basal lamina and the muscle fiber, assuming functions of maintenance of tissue homeostasis and regeneration. Muscle-specific paired box 7 (Pax7) is a hallmark of postnatal myogeneis capacity and commonly characterize SCs populations [15]. Upon injury, SCs are activated and undergo one of two faiths: differentiation into myogenic cells or "stem-like" division, maintaining the pool of available cells for intervention in future events of injury. These two courses of the SCs population relate to their Myf5 transcription factor expression: the dominant population $Pax7^+/Myf5^+$ undergoes myogenic differentiation while the minor $Pax7^+/Myf5^-$ population remains undifferentiated replenishing the SCs niche [22]. Other populations with stem cells' characteristics have been identified within the muscle tissue, such as mesangioblasts and PICs ($PW1^+/Pax7^-$ interstitial cells). Further details on intrinsic regenerative populations associated to skeletal muscle and SCs origins and dynamics can be found elsewhere [23–27].

In the final remodeling phase, the regenerated tissue matures and the formed fibrous connective tissue reorganizes and contracts [16]. This stage is highly significant for the outcome of the whole process and fine regulation of the late fibrotic events turns essential [28–30]. Especially after severe tissue loss, the fibrin matrix derived from the clotting process and inflammatory response requires remodeling into collagen type 1 network, produced by fibroblastic cells [31]. The development of definitive fibrosis at a lesion site begins at approximately 2 weeks after injury and progresses over time. Exacerbated fibrosis prejudices the repair and remodeling phases hindering muscle regeneration and full functional recovery [32].

Although fibrosis is mostly referred to as a negative aspect of the healing process, evidence suggests that a certain level of fibrosis acts as support matrix to new tissue ingrowth, promoting proper realignment of the myofibers and the myofibrils, and maintaining a degree of mechanical properties on the regenerating tissue [32]. Also, reports of "functional fibrosis" support its importance to the process, by contributing to a certain stance to the force distribution along the muscle or muscle group, preventing continued overload of the remaining skeletal muscle tissue, and contributing to functional recovery unrelated to effective skeletal muscle tissue regeneration [33].

3. MSCs' sources for skeletal muscle regeneration

In the ever-growing field of regenerative medicine and tissue engineering, stem cells are posing as one of the main characters in the most recent therapeutic strategies [17, 18].

Given the presence of resident stem and SCs within the skeletal muscle tissue, native tissue skeletal muscle-derived MSCs (MDSCs) would appear as the favorite source for regeneration therapies [34]. In response to muscle damage, the SCs population is activated by the released biomolecules and begins to proliferate and originates large numbers of muscle progenitor cells (MPCs), which will in turn contribute to skeletal muscle structure reconstruction. At the same time, not all SCs derive into MPCs but rather self-renew, contributing to the replenishment of the quiescent cells within the muscle tissue [35, 36].

Many authors have explored skeletal muscle tissue-specific cells for repair and regenerative purposes, as summarized by Koning et al. [37]. Despite of their demonstrated benefits in several settings [38], the practical use of postnatal skeletal muscle progenitors or SCs is limited due to decreased cell availability (requiring the harvest of large volumes of healthy tissue for adequate numbers) [31, 39]. The expansion of MDSCs is possible, but as it has long been known, it leads to dedifferentiation of early committed myogenic cells [40] and loss of potential. As little as 1 day in culture following isolation and sorting hinders its engraftment potential and contribution to regeneration events, hence turning it difficult to attain relevant cell numbers for implantation [34]. Confirming this loss of potential, the implantation of freshly isolated SCs in numbers as low as 250 cells outperforms the use of as many as 1.5×10^5 MPCs of first passage derived from SCs expansion [35]. This initial boost provided by seldom expanded MPCs, however, does not seem to sustain for long term effects [41].

Consequently, current research has focused on the influence of nonmuscular MSCs on promoting tissue healing and limiting fibrotic scar formation, as well as on the modulation of the inflammatory response to injury.

From the first suggestion of MSCs' potential for regenerative medicine and tissue engineering, many applications have been explored for a variety of tissues and diseases [42], including the skeletal muscle, which is the focus of this literature review. Our research group has dedicated to the development of MSC-based cellular therapies for application on several body tissues, from peripheral nerve to blood vessels and skin wounds [43–53], including for skeletal muscle volumetric loss lesions [54].

Since their first descriptions as a specific cell population in the late 1960s [55–57], knowledge on MSCs' features and potential has grown exponentially [58], as have the effective medical applications in which these are beneficial. The MSC population from the bone marrow (BM-MSCs) was the first to be characterized [55], and at the time, stem cells were thought to be exclusive to organs with fair regenerative capacity, such as the blood, intestine, bone, and skin. Nowadays, we are aware that they are present in virtually all the body tissues, in variable numbers, mostly remaining in a quiescent state until activated by significant events, ensuring a certain degree of defense against damage and disease [56, 59, 60].

The most significant features of MSCs are their clonogenic and proliferative capacities, while remaining genetically stable and in undifferentiated state, and their differentiation abilities [58], into various mesodermal, ectodermal, and endodermal cell types [61].

Through the years, significant progress has been made toward MSC characterization, and in an effort to standardize and unite the scientific community, the Mesenchymal and Tissue Stem

Cell Committee of the International Society for Cellular Therapy (ISCT) gathered a series of recommendations regarding the acceptable criteria for the definition of "mesenchymal stem cell" populations. Specifically, MSCs are determined to be characterized by (i) plastic adherent ability; (ii) absence of definitive hematopoietic lineage markers, such as CD45, CD34, CD14, CD11b, CD79-α, CD19, and class II major histocompatibility complex (MHC) molecules, specially human leukocyte antigen (HLA)-DR, and expression of nonspecific markers CD105, CD90, and CD73; and (iii) ability to differentiate into mesodermal lineage cells, osteocytes, chondrocytes, and adipocytes [62].

Another appealing point on MSC research is their immune features. Unlike terminally differentiated cells, MSCs are somehow immunologically privileged, avoiding the use of additional immunosuppressive supplements during the treatments, which are mostly (although not exclusively) deleterious for intrinsic regeneration mechanisms [42, 63]. One of the main mediators of immune responses is the HLA-II, of which MSCs present only neglectable levels, deeming them immunologically privileged [64]. This is a key point, considering the difficulty of finding matching donors among the human population and the challenges of harvesting sufficient numbers of cells from one patient upon necessity [65]. Hence, the lack of HLA-II opens the possibility of using directly obtained or banked cells from consenting healthy donors from the same species, designated as allografts [66]. Given their peculiar immune features, the xenogenic implantation of human-derived cells in appropriate nonimmunosuppressed animal models is feasible [52, 54, 67] and provides valuable information on their behavior and effect on experimental stages that more closely mimic clinical practice reality [63].

In addition, immunomodulatory actions have also been attributed to MSCs, by controlling and modifying host immune response, either locally (by blunting the tissue response at the implantation site [54]) or systemically (ameliorating signals of severe immunological disturbances, such as chronic inflammatory, autoimmune diseases or graft-versus-host-disease) [63, 68].

The bone marrow is without a doubt the most widely explored source for MSCs for therapeutic purposes. The bone marrow is harvested from the patient or consenting donor, and the adherent MSCs are isolated and expanded until desired numbers are attained for the intended application. The harvesting procedure is however highly invasive and potentially painful, motivating the search for more easily accessible sources. Furthermore, the "quality" of the isolate cells strongly depends on the age, gender, and health status of the patient or donor [65]. Adipose tissue and synovial membrane are also valid sources, and harvesting tissue for cell isolation is mostly associated to primary interventions for esthetical and/or medical reasons. [2, 39]. Other sources of MSCs are gaining ground for the minimally invasive nature of their harvest, as well as for the lesser ethical concerns surrounding their tissues of origin, namely, the stromal tissue of the umbilical cord [66] and the dental pulp [69] (Figure 1). The collection of the tissue sources for these implicates lesser ethical and technical issues since they were mostly discarded as medical waste following birth or dental procedures [66, 67, 69, 70]. Another alternative method could be the collection of MSCs from postmortem tissues. MSCs have been successfully isolated from the bone marrow, skeletal muscle, neural tissue, and dental pulp of deceased donors [71, 72]. These options, however, comply with similar if not aggravated ethical, legal, and even social and religious concerns to conventional MSC sources [71].

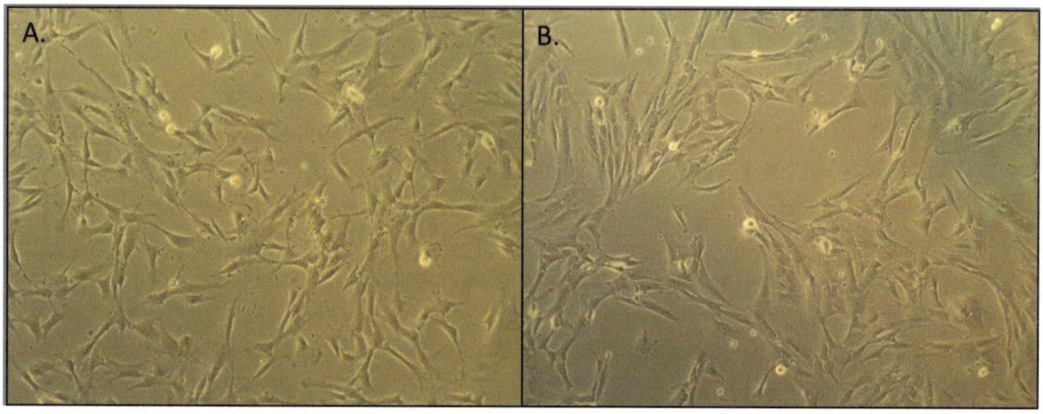

Figure 1. Morphological similarities between (A) DPSCs and (B) UC-MSCs (magnification: ×100).

Focusing on the skeletal muscle repair and regeneration, and apart from muscle-derived cells, the most popular source of MSCs is still the bone marrow. However, MSC populations are present in virtually all body tissues, and alternative sources have been proposed, such as adipose tissue, synovial membrane [2, 73, 74], dental pulp [4, 69, 75], and even umbilical cord tissue [76, 77]. Cells from these sources display comparable phenotypical features regarding their "stemness" potential (they are plastic adherent, positive for markers of the mesenchymal and negative for the hematopoietic lineage markers). As previously referred, they are similarly amenable of differentiation into mesodermal and endodermal cell lineages, including adipocytes, chondrocytes, osteoblasts, endothelial cells, and hepatocytes [61]. Nonetheless, MSCs from distinct sources are not completely identical, whether regarding phenotypical markers expression, proliferation rates, or even multilineage differentiation aptitudes [74].

3.1. Evidence of the *in vitro* myogenic differentiation potential of MSCs

MSCs from various nonmuscular sources have been demonstrated as capable of *in vitro* differentiation into myogenic lineages, through adequate stimuli, displaying phenotypical markers of native skeletal muscle cells [61]. However, nonmuscular MSCs depict hardly any spontaneous tendency toward myodifferentiation (0.001%) [6], unlike the "MSC-like" populations residing in the muscle tissue that show a degree of lineage commitment to myogenic formation, undergoing spontaneous differentiation in standard culture [31]. Not all MSC sources seem to show the same tendency or predisposition toward this differentiation pathway upon stimuli. As an example, the myogenic potential of MSCs from the synovial membrane (SM-MSCs) seems more limited than for chondrogenesis, osteogenesis, or adipogenesis since only limited number of individually expanded clones presented myotube formation capacity, suggesting subpopulations with specific tendencies for lineage commitment [73]. MSCs isolated from the umbilical cord stromal tissue display limited intrinsic tendency toward myogenic lineage differentiation, expressing diminished levels of pluripotency and specific myogenic markers involved in such process (such as Oct-4, Nanog, Pax-7, MyoD and myogenin, and M-cadherin), and they do not seem to spontaneously differentiate toward this lineage

[76]. Conversely, when cocultured with differentiating myoblasts, differentiation can be observed [76].

Cells undergoing myodifferentiation paths sequentially express characteristic transcription factors belonging to the myogenic regulatory factors (MRFs) family, such as MyoD and myosin (early a terminal differentiation markers) that are expressed with a well-defined time course depending on how far long the process has progressed [39, 78], replicating the embryogenesis of skeletal muscle tissue [15].

MSCs from the bone marrow, adipose tissue (AT-MSCs), synovial membrane, and dental pulp (DPSCs) are also capable of fusing to myoblasts in coculture systems. Although multinucleated hybrid myotubes generally appear only at low frequencies in the total population [6, 38, 74, 75] t results in the detection of muscle-specific gene expression in stromal cells, which is turned on through myogenic fusion [79].

In alternative to direct contact coculture, trans-well settings can also efficiently induce differentiation into skeletal muscle of AT-MSC subpopulations, via a fusion-independent mechanism, leading to the expression of aforementioned differentiating myotubes markers. This suggests that the differentiating myoblast can promote MSC myogenesis through secreted biomolecules that can effectively cross the trans-well filter and exert action on the MSC receptors. Nonetheless, the differentiation efficiency did not match the direct contact settings, deeming cell-to-cell direct interaction a key factor and suggesting that these two mechanisms act in complementary ways [6].

Further away from the coculture system, the supplementation of MSCs with conditioned medium (CM) from both mature muscle cells and primary precursors induced differentiation toward myogenic phenotypes [80]. CM from injured skeletal muscle also influences MSC proliferation, in a dose-dependent manner, and promotes myogenic lineage differentiation into the characteristic morphologies and transcription factors sequential expression. The medium from undamaged muscle did not elicit such responses, demonstrating that the injury event triggers the secretion of essential signaling biomolecules that modulate intervenient cells' fate *in situ* and are capable of modulating exogenous cells, such as BM-MSCs [78].

In 5-azacytidine (5-Aza)-induced differentiation, enriched umbilical cord stroma (UC-) MSCs, adherent fraction of umbilical cord blood (UCB-MSCs), periodontal ligament-derived MSCs (PDL-MSCs), SM-MSCs, AT-MSCs, BM-MSCs, and skeletal muscle-derived MSCs (SkM-MSCs) also begin displaying suggestive myoblast-like shape and fusing into multinucleated immature myofibers, expressing early muscular markers, such as Myf5 and then MyoD [2, 39, 67, 70, 77, 81]. The spontaneous twitching of multinucleated fused differentiating myotubes has also been described after 9–10 days culture [78]. Although classical myodifferentiation protocols rely on the pathway triggering by 5-Aza-induced DNA methylation, this is known to cause epigenetic changes, possibly precluding further advancements into clinical applications. In an alternative approach, differentiation can be successfully induced using a more "physiological" induction medium, composed of defined growth factors, such as bFGF, VEGF, and IGF-1, and it can successfully induce BM-MSCs into multinucleated myotube-like structures, with striated cytoplasm and replicating specific expression patterns of the myogenic pathway, similarly to other induction techniques [61]. Similar behavior is also observed in MSCs cultured in promyogenic medium containing dexamethasone and hydrocortisone [82].

3.2. Evidence on the application of MSCs for *in vivo* skeletal muscle regeneration

Further, *in vivo* applications suggest that, to variable extent, MSCs are capable of integrating host muscular tissues, being identifiable at the lesion site shortly after implantation (Figure 2) [54, 67, 75–77, 83–85]. The long time observation of administered MSCs in host muscle has also been reported, even when delivered systemically [2, 4]. Intra-arterial delivery appears more adequate for systemic MSC delivery, in detriment of intravenous routes. Keeping in mind the circulatory anatomy, this was an expected observation since venous routes implicate increased systemic dilution effects, as well as significant entrapment of cells within the pulmonary capillary bed [86].

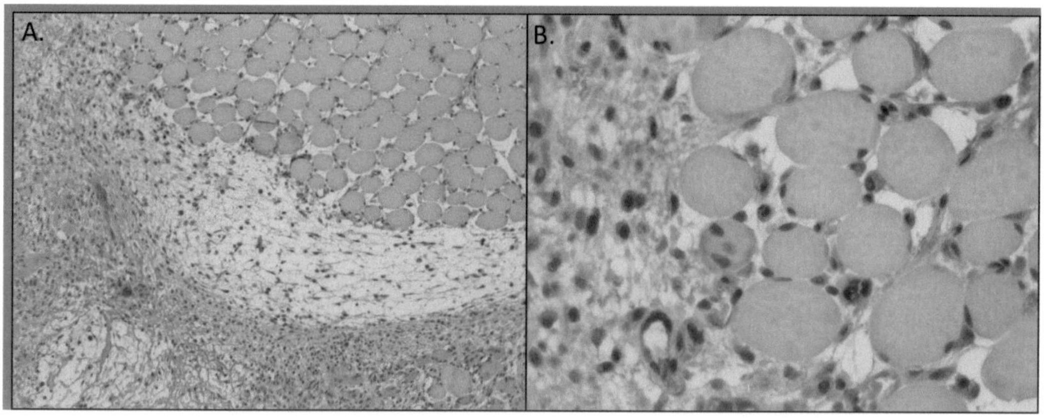

Figure 2. Human-derived UC-MSCs engrafted within injured skeletal muscle tissue 4 days after implantation. Immunohistochemistry staining for human nuclei (hNu) antigen (blue stained cells). Magnification: (A) ×100, (B) ×400.

It is set that the engraftment potential of MSCs into a damaged tissue is not absolute. The numbers or percent of cells identified at different time points following injection invariably decreases [76, 85, 87, 88], down to nearly as little as 10% of the initially delivered numbers in a couple of months [88].

MSCs positively influence recovery of chemically induced muscle damage [76, 77], nonvolumetric laceration [89] as well as in crush injuries [90, 91]. Although on occasions the delivered cells could not be identified on site as differentiated entities or fused to host cells or, when so, only to low degrees, the observed benefits further suggest that their contribution to the regeneration of skeletal muscle might rely on mechanisms other than fusion to myofibers after differentiation [89, 90].

UC-MSCs also engraft in the resident skeletal muscle tissue and are identifiable up to 14 days after administration and in some extent differentiate into cells expressing sarcomeric tropomyosine antigens [67]. When administered in undifferentiated state MSCs seem to replicate embryonic myogenesis events, as they do following *in vitro* induction [67], triggering gene expression characteristic of myogenic differentiation pathways [2].

SM-MSCs are detectable integrated within host tissue for up to 6 months after either local or systemic delivery, demonstrating their preferential homing ability to the injured tissues since greater numbers were harbored by the injured muscle, although they could seldom be identified in other body systems [2]. At longer terms after BM-MSC [88] or UC-MSC [76] administration, about 5% of myofibers were of hybrid nature and could be identified along the whole muscle length [88]. The hybrids' formations seem to be a progressive process since hybrid myofibers represent a much smaller fraction (under 1%) in the regenerating muscle at shorter time points [74]. The administration of these cells granted increased muscle mass and mature fiber formation when compared to untreated muscles [76]. AT-MSCs also contribute to enhanced regeneration, reducing fibrosis and improving histological and functional features after only 4 weeks. However MSCs alone did improve the process, their association to a biomaterial vehicle and bioactive cues further enhanced those results, as detailed later in this section [83].

In the last decade, the question whether the seldom identified donor-derived cells resulted from trans-differentiation events of MSCs into muscle cells or from fusion to host cells raised significant controversy [79]. Today, based on the acquired evidence, the scientific community tends toward the fusion theory. As mentioned earlier, MSCs have been demonstrated to successfully fuse to host cells and contribute with genetic information, leading to the expression of human-derived genes and gene products [2, 79, 88]. Fusion efficiency seems to differ among MSCs sources, and AT-MSCs appear more prone to *in vivo* formation of hybrid myotubes, surpassing BM-MSCs and SM-MSCs [74].

Delivered cells also seem to interact with the resident tissue's satellite pool. Skeletal muscle SCs delivered to a regenerating muscle effectively contribute to the replenishment of the resident satellite pool, migrating to locations far from the lesion site within the muscle. These freshly isolated native skeletal muscle stem cells contribute in high extents to the total $Pax7^{+}$ population (about 38%), unlike expanded MPCs, that only account for about as little as 12% [35], but can still be identified within the muscle tissue and in association to connective structures [31]. Nonmuscular sources seem to behave closer to MPCs in terms of niche replenishment. Based on the location of the remaining delivered cells not associated to hybrid fibers, some authors suggest that these residual donor-derived cells might have also contributed to the resident SCs pool, and that they can effectively contribute to skeletal muscle regeneration in the events of a new injury [2, 35, 76, 88]. Alike exogenous SCs and MPCs integrated into the resident quiescent pool, other sources of engrafted cells are amenable of activation upon reinjury of the muscle [2, 35] and can be reisolated [35] and also originate primary myoblast cultures *in vitro* [2]. These expanded myoblast cultures originating from human-derived isolated SM-MSCs could further engraft a second recipient's regenerating muscle [2].

As suggested by some authors, the different engraftment potential might be due to the distinct MSC sources reported [35, 74], but it may also relate to the chosen disease model: dystrophic muscle models seem to better adopt exogenous cells than non-dystrophic-induced lesion

http://dx.doi.org/10.5772/60919

models [88]. Although some disease models, especially of degenerative nature, seem to display sex-related differences, physical models of skeletal muscle lesions appear to respond similarly, at least in terms of functional recuperation. As for fibrosis development, male specimens might display a decreased tendency for the event [90]. The gender, age, and health status of the MSC donor is also a factor known to influence cell quantity, quality and general performance [65].

In line with the low engraftment potential and low contribution to *de novo* myotube formation and the solid observations that nevertheless MSCs tend to consistently lead to structural and functional improvement in skeletal muscle repair and regeneration models, the paradigm on the mode of action of MSCs started to shift. Evidence supports that their primary mechanism of action may rely on their secretory abilities rather than on their differentiation capacities, as discussed in detail further ahead [89, 90].

3.3. MSC-biomaterial systems for *in vivo* skeletal muscle regeneration

Most of the above-mentioned diseases and disease models involve severe affection of skeletal muscle tissue function (laceration or chemically induced damage) but, generally, the structural integrity of the tissue is maintained, preserving the blood and neural supply to the muscles as well as the resident SCs population [67, 88]. Although skeletal muscle detains a fair capacity for regeneration, severe injuries involving the loss of extensive volumes of muscle, termed as volumetric muscle loss (VML), mostly overwhelm this intrinsic response [12]. To date, these situations pose a relevant therapeutic challenge.

The gold standard for the surgical management is the creation of muscle flaps filling the defective area. However, these autoflaps depend on the maintenance of an adequate blood supply and involve the damage of a neighboring muscle. Therefore, the donor site morbidity and limited success of such approaches push toward the development of new treatment options [92].

The advent of tissue engineering and regenerative medicine research, focusing on both biomaterials, cells, and bioactive molecules, has boosted the search for new possibilities for the development of effective clinical treatment of affected patients [11, 92–94].

These cases are mainly related to traumatic or surgical events and result in complete tear of the myofibers or even significant loss of skeletal muscle tissue portions, VML, in large and relevant active muscle groups. Here, no support structures remain on the lesion site, nor do blood vessels, neural structures, or cell populations with capacity to repair and restore the lost tissue. The loss of 10% to 20% of the mass of a weight-bearing muscle represents a critical loss that will not fully regenerate even after long periods [8] (Figure 3).

The healing of these severe injuries can be improved by the sole administration loose of cells [34], but for the most cases, complete repair of such defects remains dependent on the ability of bridging the gap between the transected muscle segments. For VMLs, this point presents an impending challenge.

The research field of biocompatible biomaterials has opened a possible strategy to address previously irreparable lesions. These materials aim at providing a physical support to the

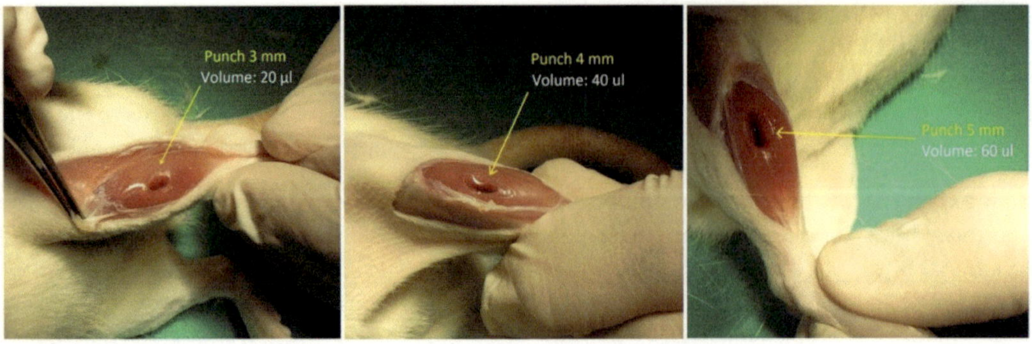

Figure 3. VML lesion model development: myectomy lesions in the tibialis anterior muscle of the rat's hindlimb. Different volumes (20–60 µl) in the defects produced by biopsy punch blades of different sizes (3–5 mm) [51].

regenerating myofibers on both ends, promoting their development and proper orientation, a key point for structural as well as functional recovery [8]. These biomaterial scaffolds are also valuable in recreating an advantageous mechanical and chemical microenvironment for the proliferation and differentiation of resident or delivered progenitors [8]. In addition, these matrices may act as (as in the case of ECM) [93] or be modified/loaded with bioactive signaling molecules participating in the repair process [83].

Here, decellularized ECMs appear as the most widely explored scaffold material for skeletal muscle tissue engineering, with commercially available products ready for clinical use and amenable of application in skeletal muscle repair and regeneration [33].

Decellularized skeletal muscle ECM has been demonstrated to adequately fill a critical muscle defect, and to benefit structural recovery. Although structural improvement was determined, functional outcomes did not significantly differ between the bridged and the unbridged control defects [8]. Other study reported functional indexes recovery comparing to the unimplanted group, but still insufficient to match undamaged muscles' response [94]. Scaling up to a larger preclinical canine model, these implants were capable of promoting endogenous progenitors migration into the regenerating area [95].

On a clinical setting application, commercially available ECM products have been successfully applied to restore chronic volumetric muscle injuries [11, 94, 96], reporting esthetic and, to some degree, functional improvements.

Besides large voids in the muscle tissue, biomaterials can also be applied in smaller defects. A gelatin-based hydrogel could enhance regeneration after laceration injury. This *in situ* forming gel was loaded with a prosurvival growth factor (bFGF), sustaining controlled release for up to 3 weeks. *In vivo*, the loaded hydrogel significantly improved contractile force of healing muscles, reduced local fibrosis, and increased multinucleated regenerating myofibers and neurovascular structures on site [83].

Stem cell implantation is a possible strategy to enhance the recovery rates whether they are delivered alone or in association to biomaterial scaffolds. Their inclusion has been demonstrated beneficial in skeletal muscle injury.

As a first approach, SCs and other muscle-derived progenitors were investigated toward the optimization of the healing process of biomaterial treated defects [41, 97].

The performance of fibrin-based [31] and hyaluronan-based hydrogels [35] was significantly improved by the association of cellular systems. MPCs were able to reduce inflammatory infiltration and scar formation and regeneration events were improved, with reinnervation and revascularization of the area as well as increasing the regenerating myofibers content. The inclusion of freshly isolated SCs instead of expanded MPCs presented even better results with functional parameters closely meeting up to controls after 6 months of recovery [35].

Given the pros and cons earlier discussed regarding native tissue-derived cells and the effects of undifferentiated nonmuscular MSCs on other lesion models, some focus was given to their potential role in cellular-biomaterial systems.

The benefits of BM-MSCs in a severe VML models through the inclusion of the cell system into a decellularized ECM frame boosted both structural and functional recovery, with increased muscle tissue, blood vessels and nervous supply ingrowth into the defect area and improved muscle functional performance, when compared to the cell-free systems [8, 98].

The potential of AT-MSCs was also tested on a previously described gelatin-based hydrogel vehicle. Although the bFGF-loaded hydrogel alone performed satisfactorily, the addition of hAT-MSCs granted further improvements. The most striking result was the reduction of fibrosis to only roughly 20% and the recovery of functional parameters reaching 89% of uninjured muscles. Indicators of regenerative events (immature myofibers, reinnervation, and neovascularization) were also significantly improved [83]. When associated to Matrigel™, early improvements were observed. However, such differences were no longer evident in a 4-week time point [87].

Besides the obvious impact on the overall regenerative milieu, the exogenous MSCs are also strong modulators of tissue reaction to biomaterials implanted within the muscle tissue.

Using a volumetric loss rat model [51], UC-MSCs demonstrated their potential in modulating early inflammatory responses to a gelatin/thrombin-based matrix [52]. It was further confirmed in response to other biomaterial systems in terms of both the inflammatory response and the collagen type I deposition. The results from the sole application of a good vehicle (fibrin-based) were further improved, and the reaction to vehicles deemed less adequate (gelatin/thrombin-matrix and hyaluronan/alginate-based) was attenuated by the presence of UC-MSCs [54].

As described in an earlier section, one of the challenges faced on the use of cell systems is the success in engrafting the lesion site. The preconditioning of BM-MSCs to the myogenic lineage seems to improve integration in host muscles [61], but the pretreatment of UC-MSCs with SDF-1 does not seem to affect engraftment efficiency, despite the fact that SDF-1 is a known promoter of migration of transplanted and host cells to active lesion sites [76]. By contrast, the combination of bFGF, VEGF, and IGF-1 positively impacted engraftment, with increased donor cells' numbers identified forming mature hybrid myotubes [61].

The delivery of the intended MSCs within a scaffold/vehicle also contributes to prolonged maintenance on site, increasing engrafted cells number comparing to loosely delivered cells

[83]. Hence, the mode of delivery of cells or other regenerative cues is of vital importance [54]. Loosely delivered cells tend to show poor survival and engraftment and inadequate interaction with the host tissue. These drawbacks seem to be counteracted by their association to delivery vehicles [35] that also seem to positively influence cell survival and myodifferentiation, as well as neovascularization of the lesion sites when implanted [83].

In summary, both biomaterials and cells alone can aid the healing process, but their association seems to boost their individual actions. Cells help in functionalizing biomaterials while biomaterial provided beneficial microenvironments for the survival and action of the encapsulated MSCs.

From the currently available data, biomaterials alone are capable of providing fair benefits to volumetric lesions, but longer periods of recovery might be required (over 6 months). The coordination of those with cellular systems is likely to speed up the process, providing evidence of functional recovery earlier after the treatments [97].

The type and the magnitude of the contribution of the seeded population to the regenerative process also seem to relate to their differentiation state. Cells closer to undifferentiated state seem to elicit boosted initial responses, accelerating the onset of the process [87, 97]. By contrast, specialized cells tend to provide a more gradual but sustained response. The combination of the two populations may provide the key for additive effects and magnified recoveries [41].

One of the most striking observations is that the application of lineage committed or undifferentiated cells correlates to increased vascularization (and also innervation) at interface and core areas of implanted materials, which is known to be a vital factor for cell survival and function and tissue regrowth into volumetric matrices [98], since three-dimensional scaffolds easily exceed the diffusion capacity for nutrient and other essential components toward the inner parts of the constructs [99]. Seeding with potentially vasculogenic cells and/or prevascularization of constructs via *in vitro* culture seem viable options for improved results upon implantation into critical defects [100, 101]. The association of seeding strategies to surgical vascularization (i.e., connection to hosts vessels) is also described as boosting regeneration and recovery from large defects [102].

The timing of administration might also impact on MSCs engraftment and function. Most authors describe the existence of delivered cells for weeks following implantation [67, 76, 77, 83], when delivered up to 24 hours after injury. This topic remains highly debatable. The delivery of MSCs into a crush injury model either immediately or 1 week after the event did not lead to significant differences in functional recovery, indicating the possibility of a fairly large time window for the application of these therapies [84]. However, others report that the delivery of MSCs 1 week after injury (in attempt to escape initial inflammatory reaction) seems to impair their engraftment since no trace was detected after only 2 weeks. Most surprisingly, the cells were delivered on a hydrogel vehicle [87] what had been described to positively influence the permanency of delivered cells at the lesion site [83]. It might suggest that the early inflammatory microenvironment modulates MSCs function and maintenance on site or that their modulatory effects on the inflammatory milieu affect their engraftment.

Another critical factor under investigation is the most adequate number of MSCs to be delivered to the defect site, associated or not to a biomaterial vehicle. Winkler and colleagues

http://dx.doi.org/10.5772/60919

demonstrated a dose-dependent response to BM-MSCs administration in functional recovery from 0.1 to 10 million cells administered to a crush injury site [103]. Irrespectively, the number of cells administered regardless of the disease model or cell source or vehicle depicts no consensus among research groups, ranging from few thousands to several millions [52, 54, 67, 76, 77, 88, 98, 103].

Cell Source	Lesion Model	Delivery Mode/ Vehicle	Reference
hBM-MSCs	CTX TA mice	local injection	[88, 74, 79, 61]
rBM-MSCs	VML GTN rat	ECM	[98]
rBM-MSCs	CR SL rat	local injection	[103, 90, 91, 84, 85]
rBM-MSCs	CR SL rat	systemic delivery	[86]
hUC-MSCs	VML TA rat	thrombin-based matrix	[51, 52, 54]
hUC-MSCs	VML TA rat	fibrin-based matrix	[54]
hUC-MSCs	VML TA rat	hyaluronan/ alginate-based hydrogel	[54]
hUC-MSCs (enriched)	BVC TA rat	local injection	[67]
hUC-MSCs	CTX GTN mice	local injection	[76]
hAT-MSCs (+ bFGF)	LAC GTN mice	hyaluronic acid- based hydrogel	[83]
rAT-MSCs	LAC SL rat	Matrigel ™	[87]
hAT-MSCs	CTX TA mice	local injection	[74]
hUCB-MSCs	CTX GTN mice	local injection	[77]
hSM-MSCs	CTX TA mice	local injection systemic delivery	[2]
hSM-MSCs	CTX TA mice	local injection	[74]
hDPSCs	CTX TA mice	local injection	[75]

Table 1. Examples of non-muscular MSCs sources for *in vivo* skeletal muscle repair and regeneration; r: rat or mice derived cells; h: human derived cells; CTX: Cardiotoxin chemical model; VML: Volumetric Muscle Loss model; CR: Crush Injury model; BVC: Bupivacaine chemical model; LAC: Laceration injury model; TA: *Tibialis anterior* muscle; GTN, *Gastrocnemius* muscle; SL: *Soleus* muscle; ECM: Extracellular Matrix.

Since the core topic of this section is *in vivo* skeletal regeneration of acquired lesions models, an important remark must be made. Although the general intrinsic regenerative response of the skeletal muscle tissue to an injury event (approached in detail earlier in this chapter) follows a mostly constant sequence of events, the source and the magnitude of the induced injury lead to distinct native healing efficiencies and consequently distinct responses to implemented therapies [27, 104]. It results into an additional challenge when trying to understand the true magnitude of effects, especially in attempts of comparative analysis on the available literature.

Besides the selected defect model, the animal species/strain also assumes relevance. Different animal strains within the same species depict distinct tissue and systemic response profiles to a similar injury [105]. Also, the use of immunocompromised [31] or non-immunocompromised [54] animals may also contribute to the inflammatory responses obtained when biomaterials and xenogenic cell sources are applied.

Other consideration possibly precluding the translation of developed therapies to the clinical practice is the character of the lesion site. In research scenarios, therapies are applied to recently injured sites. However, on a clinical setting, the most expectable situation is a chronic irreparable wound that underwent multiple surgical repair attempts though the course of several months or years [11, 94, 96]. Hence, as pointed out by Vigodarzere and Mantero [92], the homeostasis of such extensively injured and remodeled sites is significantly distinct from freshly induced insults; thus, the predictive value of the currently used animal models turns questionable.

4. Secretory potential of MSCs and impact on skeletal muscle regeneration

As evidenced by several authors, the beneficial action of MSCs on regenerating skeletal muscle might not solely depend on their differentiation capabilities, especially in nondegenerative lesion models, where their engraftment capacities seem fairly limited [76, 84, 88–90, 103]. Other proposed action mechanisms involve the secretion power of those cells [42, 76] since relevant growth factors and cytokines have been identified in various MSCs' sources secretome profiles [106, 107].

4.1. The role of growth factors and cytokines in skeletal muscle regeneration

The basis for this approach rests on the evidence that specific growth factors influence skeletal muscle regenerative response [108, 109]. In injured tissues, these factors are secreted into the surrounding microenvironment, exerting effect on, as an example, quiescent progenitor cells or delivered MSCs [78]. Thus, upon injury, the skeletal muscle itself releases a cascade of modulatory and signaling biomolecules, aiming at the recruitment and activation of essential characters to the regenerative process and triggering cell-type-specific programs [15]. These secretory capacities can inclusively modulate *in vitro* cell cultures and induce undifferentiated MSCs toward myoregeneration. In contrast, undamaged muscle seems to remain in a quiescent state, devoid of active stimulatory or differentiating factors [78].

Relevant growth factors are secreted by the remaining tissue but also by the invading immune cells participating in the intrinsic inflammatory response. Some of these molecules act as chemoattractant to additional inflammatory infiltration to the lesion site, such as MCP-1, IL-17, TNF-α, and TGF-β, among many others (a comprehensive table on the normalized nomenclature for some growth factors and cytokines is available as supplementary material on [107]) [110]. Macrophages are a grand character of skeletal muscle inflammatory response and accompany the full process of recovery [111], modulating their phenotype and secretory abilities and interaction with neighboring cells. They are chemoattracted to the site by molecules deriving from damaged muscle cells and other populations, such as neutrophils. They primarily secrete TNF-α and INF-γ then shift to increased levels of IL-4 and IL-10, promoting initial SCs division in undifferentiated state and, later on, their differentiation toward myogenesis [110]. IL-4 is also actively secreted by eosinophils active in the early stages of response to muscle insult. At this time point, this cytokine is essential to the activation and action of resident cell populations fibro/adipogenic precursors, promoting their proliferation while inhibiting differentiation into adipogenic lineages, contributing to the formation of essential support structures to aid myotube regeneration and to further secretion of bioactive factors [112].

Parallel to the types of growth factors and cytokines involved, it is essential to bear in mind that release/delivery dynamics is also of vital importance [27]. The intrinsic regenerative mechanisms following skeletal muscle damage does rely on the sequential and coordinated interaction of molecules [113], and the key to the development of improved strategies might come from contemplating and replicating these facts. Therefore, besides the growth factors content in a lesion site, the strict patterns of interaction between those play a crucial role in the outcome of the regenerative process. For example, HGF and bFGF activity after crush injury increases during the early regeneration period (first week), while TGF-β3 only significantly increased later in the process (after 12 days postcrush) [114].

HGF is a potent mitogen for quiescent SCs, inducing their activation and increasing the numbers of proliferating MPCs while preventing their differentiation [115, 116]. The effects of HGF in SCs quiescence appears to be the work of a concentration-dependent negative-feedback mechanism, promoting activation and proliferation at low concentrations, while rebooting SCs to quiescence and promoting muscle-specific proteins expression in increasing concentrations [117]. It is present in the undamaged muscle and is released upon injury [118], mainly of physical/mechanical nature [117], and it is also released from other organs, such as the liver and spleen, acting on skeletal muscle tissue in an endocrine way [119]. Its effects are observed in a restricted time window, peaking for the first days following injury and then decreasing. Given its inhibitory effect in myodifferentiation, its role in later stages of the regenerative process turns deleterious [116], if low expression is maintained [117].

Basic-FGF and IGF-I have also been reported to positively influence muscular cell populations in both *in vitro* and *in vivo* settings. In myoblast cultures, bFGF, IGF-I, and NGF effectively promoted myoblast proliferation and fusion into multinucleated myofibers, while other factors such as PDGF-AA, EGF, TGF-α, and TGF-βs seemed detrimental in that specific setting [120]. *In vivo* application into a laceration injury confirmed the beneficial effects of bFGF and IGF-I [120]. Basic-FGF sustained release at a lesion site (i.e., via hydrogel delivery) also elicits increased revascularization and reinnervation of the regenerated tissue, reducing the development of fibrosis [83]. IGFs are associated to both muscle cell proliferation and differentiation and play a key role in muscle regeneration and hypertrophy, with different isoforms affecting different stages of the process. IGF-IEc/MGF is expressed early in events, associated to SCs and proliferating myoblasts, while IGF-Ia and IGF-II expression occurs later in myogenic differentiation and muscle fiber formation [121]. IGF-I also modulates the local inflammatory response by down-regulating inflammatory cytokines on site, and thus limiting fibrosis development [122]. The combination of anti-fibrotic agents to IGF-I administration exerts additive effects on muscle regeneration [32]. Sustained IGF-I delivery enhances ischemic muscle fiber regeneration, and beneficial effects are potentiated by combination with other growth factors, such as VEGF, resulting in synchronized angiogenesis, reinnervation, and myogenesis [123].

Other members of the FGF family interfere with skeletal muscle regeneration. FGF-6 is deemed muscle specific and is up-regulated during regenerative events, and its absence has been reported to relate to regenerative defects [124]. High concentrations of FGF-6 stimulate the proliferation of the myogenic stem cells, while while lower concentrations regulate muscle differentiation. It is also a determining factor for skeletal muscles' fiber type content [125, 126].

Vascular endothelial growth factor (VEGF) is also an important factor in muscle regeneration. In damaged tissue, VEGF and its receptors are detected in SC and in regenerating muscle fibers, as well as in cultured SC and myoblasts. VEGF acts by stimulating myoblast migration and survival, preventing apoptosis, and promoting myogenic cell growth. Furthermore, VEGF may have a relevant role in the homing of circulating progenitor cells to specific muscle location and/or in regulating the SC pool [127]. The local administration of VEGF has also been associated with reduced scaring and improved muscle regeneration and strength recovery after acute trauma [128]. Sustained VEGF delivery promotes neo-angiogenesis and tissue perfusion recovery, as well as conferring protection from hypoxia and tissue necrosis in ischemic limbs [123], but it may derive into aberrant ECM deposition and undesired fibrosis [129].

Granulocyte colony-stimulating factor (G-CSF) also exerts beneficial effects in skeletal muscle healing, promoting both structural and functional recovery in damaged muscles [130, 131], and is a crucial factor for skeletal muscle development [130]. It promotes myoblasts proliferation *in vitro*, in a dose-dependent manner, and although it is a recruiting factor for hematopoietic stem cells from the bone marrow, it does not influence the recruitment of BM-MSCs for skeletal muscle regeneration purposes [130].

As mentioned before, not all bioactive molecules and interactions have solely positive effects. Increased TGF-β1 levels are observed at injury sites [132]. This particular growth factor is stimulatory to collagen and ECM deposition that can be detrimental to the skeletal muscle regeneration process, contributing to exacerbated fibrosis and loss of contractile properties [32]. When TGF-β1 activity is inhibited by the action of decorin, regeneration indexes significantly improve and fibrosis development decreases by 50% in laceration injuries [133], conveying toward *in vitro* observations [132]. TGF-β1 acts on myoblasts, leading to the overexpression of fibrosis-related proteins and the down-regulation of myogenic proteins (desmin, MyoD, and myogenin). Furthermore, TGF-β1 released by injured muscle stimulates autocrine expression on surrounding myoblasts and inflammatory cells, amplifying its local fibrogenic effects [132].

These and other growth factors and cytokines are well known to guide and modulate tissue response to damage, and their coordinated actions are essential for the timely activation of myogenic cells, revascularization, and reinnervation of the lesion site and ECM deposition and remodeling.

http://dx.doi.org/10.5772/60919

4.2. MSCs secretome and effects on *in vivo* skeletal muscle regeneration

Most of MSCs secretome components are described to exert regulatory functions in both autocrine and paracrine ways [134], and interact both directly and indirectly with other cells, by triggering direct intracellular signaling pathways or by activating molecules production and release by other targeted cell types [42]. These bioactive molecules are deemed to benefit repair and regeneration processes mostly by inhibiting apoptosis and limiting the extent/propagation of injury, by diminishing fibrotic tissue development, by stimulating angiogenesis and revascularization of the regenerating tissue, and by activating/boosting intrinsic tissue-specific stem cell pools [63].

As disclosed in a previous section, the observations of positive effects upon MSCs application regardless of their presence as differentiated entities on site strongly support the assumption that their actions may alternatively depend on their capacity to produce and secrete compounds when in undifferentiated state [41, 89, 90]. This theory is also supported by the fact that exogenous MSCs mostly position themselves in close vicinity to regenerating myofibers, in native SCs/PICs-like locations, providing controlled release of such components [76]. Caplan and Dennis quite accurately described MSCs as "multi-drug delivery vehicles that are injury-site sensitive and/or responsive" [42], also referring to MSCs "homing" capacity (i.e., their ability to respond to signaling chemokines and preferentially migrate and attach close to lesion sites). Hence, the combination of molecules secreted by MSCs gain interest as modulators of inflammatory, fibrotic, and regenerating events [54].

Since the 1990s, considerable efforts have been made toward the comprehension of the secretion potential of MSCs derived from various tissues and exhaustive studies have focused on the detailed composition of their secretome [107, 135–137] and their actions and functions on the modulation of inflammatory and regenerative events, as thoroughly revised in [42, 63, 134, 138–142].

Briefly, and according to their prospective effects on regenerative processes in general, these factors and chemokines can classically be classified as anti-apoptotic, immunomodulatory, anti-scaring, supportive, angiogenic, and chemoattractant [138]. Factors including HGF, IGFs, FGFs, CSFs, PDGFs, and TGFs as well as cytokines such as IL-6, IL-8, and IL-10 have been identified in different magnitudes in MSCs culture supernatants [106, 107, 142]. Other performers in the skeletal muscle regenerative process seem to be absent, such as IL-4 [107].

The array of secreted molecules is related to the microenvironment accommodating the active MSCs, displaying consistent patterns of secretions in response to their local microenvironment, as well as to their functional status [42]. This responsiveness of MSCs to a variety of microenvironmental cues can be availed as to enhance their therapeutic potential from the amount of secreted factors [134] up to incrementing the engraftment success when implanted at a lesion site [61]. Inflammatory cues alter the expression patterns of MSCs, resulting in increased secretion of selected growth factors and other cytokines [135, 142]. MSCs can be exposed to controlled stimuli before application, such as hypoxia and mechanical stimulation, leading increased expression of growth factors such as bFGF, IGF-I, HGF and, with particular emphasis, VEGF [137]. These observations are of particular interest since severely damaged muscles present hypoxic milieus due to the impairment or loss of blood supply.

Nevertheless, most of the knowledge available on the MSC secretome derives from *in vitro* settings, and despite studies focusing on the effects of biomolecules on its profile, it hardly replicates the exact inflammatory scenario within a lesion.

Concurrently to the implantation of undifferentiated MSCs, it has also been hypothesized that the application of secretion products alone (termed as conditioned medium [CM]) display similar if not improved effects on skeletal tissue regeneration (Figure 4) [54], as it does in other damaged or degenerated tissues, such as the central nervous system [141, 143, 144].

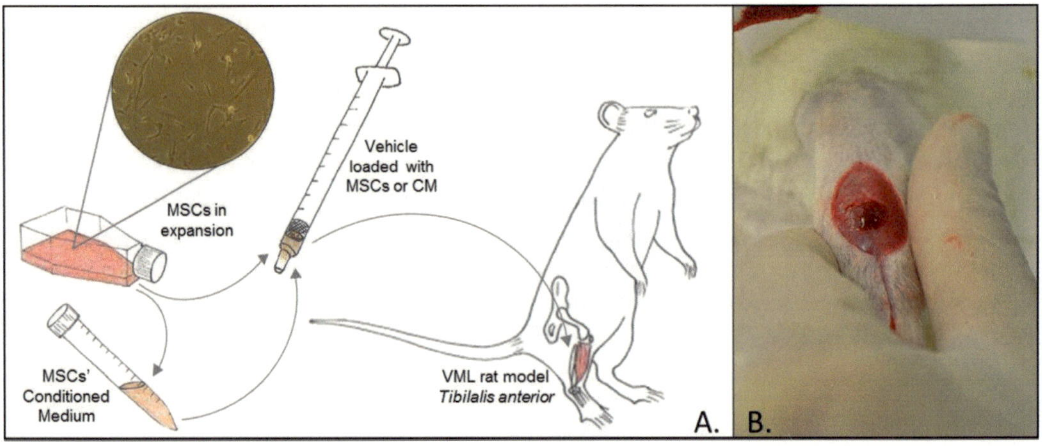

Figure 4. Schematic on the application of MSCs or their CM on hydrogel vehicles for the regeneration of critical muscle defects in a rat model (A) as described in [54]; VML lesion filled with loaded vehicle.

When comparing the skeletal muscle inflammatory response to implanted biomaterials of a severe tissue loss model, the association of either undifferentiated Wharton's jelly MSCs or CM obtained from their *in vitro* culture seems to consistently blunt the inflammatory response at early to medium stage of events (day 15 postlesion) [54]. Besides the reduced afflux of inflammatory cells to the implant sites on these early stages, we have also observed that the presence of MSCs resulted in accelerated progression to chronic inflammatory stages, that would resolve faster than unloaded subcutaneous implants (unpublished data). Therefore, although MSCs secrete a wide range of chemoattractive chemokines, their coordinate effects result in effective immunomodulation and in the control and containment of the local inflammatory response. Similarly, the addition of CM from UC-MSCs significantly improved the local response to different biomaterials as well as promoted improved muscle regeneration, revascularization, and reinnervation at the site of severe VML [54]. These results are supported by the detection of fair levels of immunomodulatory cytokines, such as IL-10 and MCP-1, as well as proregenerative growth factors, such as bFGF and VEGF [107]. Recently, we have observed similar behavior regarding the applications of DPSCs' CM in the same VML model. It similarly reduced the early to medium stages of inflammatory response to implanted biomaterial, but slight differences were observed. These differences might be attributed to slight differences in the secretome profile of UC-MSCs and DPSCs, and research is ongoing on this topic (unpublished data).

5. Conclusions and final remarks

The following are brief conclusions on the topics discussed along this chapter:

The skeletal muscle is frequently exposed to severe trauma that overwhelms its intrinsic healing mechanisms. To date, conservative and surgical treatment options often fail to restore the structure and function of the affected muscle.

The expansion of the regenerative medicine research field enlightened scientific community on some possible strategies to improve those clinical outcomes. MSCs appear as a promising source for the development of cellular therapies for skeletal muscle and other body systems. Significant achievements have been made toward their isolation from viable tissue sources, with sources like the umbilical cord or adipose tissue gaining ground over the classical bone marrow.

The recognition of nonmuscular MSCs potential for skeletal muscle regeneration lays on the observations that they can (i) assume skeletal muscle cells phenotypes (differentiate) and (ii) fuse to native muscle cells, that (iii) they can integrate living host tissues as differentiated and undifferentiated entities, and finally (iv) that they secrete a wide range of bioactive molecules with impact on the skeletal muscle regeneration milieu.

There is still great ground to cover in search for definitive therapies, but great promise holds on the development and refinement of tissue engineering strategies, combining the use of structural and active biomaterials, nonmuscular MSCs, and their secretion products in order to aid and guide the body's efforts to heal severe volumetric muscle lesion, aiming at the full recovery of the muscles' structure and function that greatly affect patients quality of live and well-being.

Acknowledgements

This research was supported by the System of Incentives for Research and Technological development of QREN in the scope of project no. 38853/2013—DEXGELERATION: "Soluções avançadas de regeneração óssea com base em hidrogéis de dextrino"—by the European Community FEDER fund through ON2—O Novo Norte—North Portugal Regional Operational Program 2007–2013, by project no. 34128—BEPIM II: "Microdispositivos biomédicos com capacidade osteointegrativa por μPIM" —BEPIM II, funded by AdI, and by the program COMPETE—Programa Operacional Factores de Competitividade, Project Pest-OE/AGR/UI0211/2011, by the FCT (the Portuguese Foundation for Science and Technology) project PTDC/CVT/103081/2008, and CDRsp's Strategic Project—UI-4044-2011-2012 (Pest-OE/EME/

UI4044/2011) funding from FCT. AR Caseiro would also like to acknowledge the FCT for the PhD grant SFRH/BD/101174/2014.

Author details

A.R. Caseiro[1,2,3,4], T. Pereira[1,2], P.J. Bártolo[3,5], J.D. Santos[4], A.L. Luís[1,2] and A. C. Maurício[1,2*]

*Address all correspondence to: ana.colette@hotmail.com, acmauricio@icbas.up.pt

1 Departamento de Clínicas Veterinárias, Instituto de Ciências Biomédicas de Abel Salazar (ICBAS), Universidade do Porto (UP), Rua de Jorge Viterbo Ferreira, Porto, Portugal

2 Centro de Estudos de Ciência Animal (CECA), Instituto de Ciências e Tecnologias Agrárias e Agro-Alimentares (ICETA), Rua D. Manuel II, Apartado Porto, Portugal

3 CDRsp-Centro para o Desenvolvimento Rápido e Sustentado de Produto, Instituto Politécnico de Leiria, Rua de Portugal, Zona Industrial, Marinha Grande, Portugal

4 School of Mechanical, Aerospace and Civil Engineering & Manchester Institute of Biotechnology, University of Manchester, Manchester, UK

5 CEMUC, Departamento de Engenharia Metalúrgica e Materiais, Faculdade de Engenharia, Universidade do Porto, Rua Dr. Roberto Frias, Porto, Portugal

References

[1] Sicari BM, Dearth CL, Badylak SF. Tissue engineering and regenerative medicine approaches to enhance the functional response to skeletal muscle injury. Anatomical record (Hoboken, NJ: 2007). 2014;297(1):51–64.

[2] De Bari C, Dell'Accio F, Vandenabeele F, Vermeesch JR, Raymackers JM, Luyten FP. Skeletal muscle repair by adult human mesenchymal stem cells from synovial membrane. Journal of Cell Biology. 2003;160(6):909–18.

[3] Corti S, Strazzer S, Del Bo R, Salani S, Bossolasco P, Fortunato F, et al. A Subpopulation of murine bone marrow cells fully differentiates along the myogenic pathway and participates in muscle repair in the mdx dystrophic mouse. Experimental Cell Research. 2002;277(1):74–85.

[4] Kerkis I, Ambrosio CE, Kerkis A, Martins DS, Zucconi E, Fonseca SA, et al. Early transplantation of human immature dental pulp stem cells from baby teeth to golden retriever muscular dystrophy (GRMD) dogs: local or systemic? Journal of Translational Medicine. 2008;6:35.

[5] Goncalves MA, de Vries AA, Holkers M, van de Watering MJ, van der Velde I, van Nierop GP, et al. Human mesenchymal stem cells ectopically expressing full-length dystrophin can complement Duchenne muscular dystrophy myotubes by cell fusion. Human Molecular Genetics. 2006;15(2):213–21.

[6] Di Rocco G, Iachininoto MG, Tritarelli A, Straino S, Zacheo A, Germani A, et al. Myogenic potential of adipose-tissue-derived cells. Journal of Cell Science. 2006;119(14): 2945–52.

[7] Goudenege S, Pisani DF, Wdziekonski B, Di Santo JP, Bagnis C, Dani C, et al. Enhancement of myogenic and muscle repair capacities of human adipose-derived stem cells with forced expression of MyoD. Molecular Therapy. 2009;17(6):1064–72.

[8] Merritt EK, Hammers DW, Tierney M, Suggs LJ, Walters TJ, Farrar RP. Functional assessment of skeletal muscle regeneration utilizing homologous extracellular matrix as scaffolding. Tissue Engineering Part A. 2010;16(4):1395–405.

[9] Gates C, Huard J. Management of skeletal muscle injuries in military personnel. Operative Techniques in Sports Medicine. 2005;13(4):247–56.

[10] Järvinen TAH, Järvinen TLN, Kääriäinen M, Äärimaa V, Vaittinen S, Kalimo H, et al. Muscle injuries: optimising recovery. Best Practice and Research Clinical Rheumatology. 2007;21(2):317–31.

[11] Mase Jr VJ, Hsu JR, Wolf SE, Wenke JC, Baer DG, Owens J, et al. Clinical application of an acellular biologic scaffold for surgical repair of a large, traumatic quadriceps femoris muscle defect. Orthopedics. 2010;33(7):511.

[12] Grogan BF, Hsu JR, Consortium STR. Volumetric muscle loss. Journal of the American Academy of Orthopaedic Surgeons. 2011;19(Suppl 1):S35-S7.

[13] Jarvinen M, Aho AJ, Lehto M, Toivonen H. Age dependent repair of muscle rupture. A histological and microangiographical study in rats. Acta Orthopaedica Scandinavica. 1983;54(1):64–74.

[14] Gopinath SD, Rando TA. Stem cell review series: aging of the skeletal muscle stem cell niche. Aging Cell. 2008;7(4):590–8.

[15] Giordani L, Puri PL. Epigenetic control of skeletal muscle regeneration: integrating genetic determinants and environmental changes. FEBS Journal. 2013;280(17):4014–25.

[16] Järvinen TAH, Järvinen TLN, Kääriäinen M, Kalimo H, Järvinen M. Muscle injuries. American Journal of Sports Medicine. 2005;33(5):745–64.

[17] MacLean S, Khan W, Malik A, Anand S, Snow M. The potential of stem cells in the treatment of skeletal muscle injury and disease. Stem Cells International. 2011;2012.

[18] Tedesco FS, Dellavalle A, Diaz-Manera J, Messina G, Cossu G. Repairing skeletal muscle: regenerative potential of skeletal muscle stem cells. Journal of Clinical Investigation. 2010;120(1):11.

[19] Borrione P, Di Gianfrancesco A, Pereira MT, Pigozzi F. Platelet-rich plasma in muscle healing. American Journal of Physical Medicine and Rehabilitation. 2010;89(10):854–61.

[20] Tidball JG. Inflammatory processes in muscle injury and repair. American Journal of Physiology—Regulatory, Integrative and Comparative Physiology. 2005;288(2):R345–53.

[21] Mauro A. Satellite cell of skeletal muscle fibers. Journal of Biophysical and Biochemical Cytology. 1961;9(2):493–5.

[22] Kuang S, Kuroda K, Le Grand F, Rudnicki MA. Asymmetric self-renewal and commitment of satellite stem cells in muscle. Cell. 2007;129(5):999–1010.

[23] Mitchell KJ, Pannerec A, Cadot B, Parlakian A, Besson V, Gomes ER, et al. Identification and characterization of a non-satellite cell muscle resident progenitor during postnatal development. Nature Cell Biology. 2010;12(3):257–66.

[24] Pannerec A, Formicola L, Besson V, Marazzi G, Sassoon DA. Defining skeletal muscle resident progenitors and their cell fate potentials. Development. 2013;140(14): 2879–91.

[25] Pannerec A, Marazzi G, Sassoon D. Stem cells in the hood: the skeletal muscle niche. Trends in Molecular Medicine. 2012;18(10):599–606.

[26] Pannerec A, Sassoon D. PICs (PW1+ interstitial cells), a new muscle stem cell population. Médecine/Science (Paris). 2010;26(10):797–800.

[27] Charge SBP, Rudnicki MA. Cellular and molecular regulation of muscle regeneration. Physiological Reviews. 2004;84(1):209–38.

[28] Serrano AL, Mann CJ, Vidal B, Ardite E, Perdiguero E, Munoz-Canoves P. Cellular and molecular mechanisms regulating fibrosis in skeletal muscle repair and disease. Current Topics in Developmental Biology. 2011;96:167–201.

[29] Serrano AL, Munoz-Canoves P. Regulation and dysregulation of fibrosis in skeletal muscle. Experimental Cell Research. 2010;316(18):3050–8.

[30] Mann CJ, Perdiguero E, Kharraz Y, Aguilar S, Pessina P, Serrano AL, et al. Aberrant repair and fibrosis development in skeletal muscle. Skeletal Muscle. 2011;1(1):21.

[31] Page RL, Malcuit C, Vilner L, Vojtic I, Shaw S, Hedblom E, et al. Restoration of skeletal muscle defects with adult human cells delivered on fibrin microthreads. Tissue Engineering Part A. 2011;17(21–22):2629–40.

[32] Sato K, Li Y, Foster W, Fukushima K, Badlani N, Adachi N, et al. Improvement of muscle healing through enhancement of muscle regeneration and prevention of fibrosis. Muscle Nerve. 2003;28(3):365–72.

[33] Corona BT, Wu X, Ward CL, McDaniel JS, Rathbone CR, Walters TJ. The promotion of a functional fibrosis in skeletal muscle with volumetric muscle loss injury following the transplantation of muscle-ECM. Biomaterials. 2013;34(13):3324–35.

[34] Tamaki T, Uchiyama Y, Okada Y, Ishikawa T, Sato M, Akatsuka A, et al. Functional recovery of damaged skeletal muscle through synchronized vasculogenesis, myogenesis, and neurogenesis by muscle-derived stem cells. Circulation. 2005;112(18):2857–66.

[35] Rossi CA, Flaibani M, Blaauw B, Pozzobon M, Figallo E, Reggiani C, et al. In vivo tissue engineering of functional skeletal muscle by freshly isolated satellite cells embedded in a photopolymerizable hydrogel. FASEB Journal. 2011;25(7):2296–304.

[36] Collins CA, Olsen I, Zammit PS, Heslop L, Petrie A, Partridge TA, et al. Stem cell function, self-renewal, and behavioral heterogeneity of cells from the adult muscle satellite cell niche. Cell. 2005;122(2):289–301.

[37] Koning M, Harmsen MC, van Luyn MJA, Werker PMN. Current opportunities and challenges in skeletal muscle tissue engineering. Journal of Tissue Engineering and Regenerative Medicine. 2009;3(6):407–15.

[38] Beier JP, Bitto FF, Lange C, Klumpp D, Arkudas A, Bleiziffer O, et al. Myogenic differentiation of mesenchymal stem cells co-cultured with primary myoblasts. Cell Biology International. 2011;35(4):397–406.

[39] Mizuno H, Zuk PA, Zhu M, Lorenz PH, Benhaim P, Hedrick MH. Myogenic differentiation by human processed lipoaspirate cells. Plastic and Reconstructive Surgery. 2002;109(1):199–209.

[40] Yaffe D. Retention of differentiation potentialities during prolonged cultivation of myogenic cells. Proceedings of the National Academy of Sciences of the United States of America. 1968;61(2):477.

[41] Corona BT, Machingal MA, Criswell T, Vadhavkar M, Dannahower AC, Bergman C, et al. Further development of a tissue engineered muscle repair construct in vitro for enhanced functional recovery following implantation in vivo in a murine model of volumetric muscle loss injury. Tissue Engineering Part A. 2012;18(11–12):1213–28.

[42] Caplan AI, Dennis JE. Mesenchymal stem cells as trophic mediators. Journal of Cellular Biochemistry. 2006;98(5):1076–84.

[43] Alexandre N, Ribeiro J, Gartner A, Pereira T, Amorim I, Fragoso J, et al. Biocompatibility and hemocompatibility of polyvinyl alcohol hydrogel used for vascular grafting- In vitro and in vivo studies. Journal of Biomedical Materials Research Part A. 2014;102(12):4262–75.

[44] Amado S, Rodrigues JM, Luis AL, Armada-da-Silva PA, Vieira M, Gartner A, et al. Effects of collagen membranes enriched with in vitro-differentiated N1E-115 cells on rat sciatic nerve regeneration after end-to-end repair. Journal of NeuroEngineering and Rehabilitation. 2010;7:7.

[45] Gärtner A, Pereira T, Armada-da-Silva P, Amado S, Veloso A, Amorim I, et al. Effects of umbilical cord tissue mesenchymal stem cells (UCX®) on rat sciatic nerve regeneration after neurotmesis injuries. Journal of Stem Cells and Regenerative Medicine. 2014;Vol.X(1):JSRM Code: 010010300001EPA010214.

[46] Gärtner A, Pereira T, Armada-da-Silva PA, Amorim I, Gomes R, Ribeiro J, et al. Use of poly(DL-lactide-ε-caprolactone) membranes and mesenchymal stem cells from the Wharton's jelly of the umbilical cord for promoting nerve regeneration in axonotmesis: in vitro and in vivo analysis. Differentiation. 2012;84(5):355–65.

[47] Gärtner A, Pereira T, Simões MJ, Armada-da-Silva P, França ML, Sousa R, et al. Use of hybrid chitosan membranes and MSC cells for promoting nerve regeneration in an axonotmesis rat model. Neural Regeneration Research. 2012;7(29):2247–58.

[48] Luis AL, Rodrigues JM, Geuna S, Amado S, Shirosaki Y, Lee JM, et al. Use of PLGA 90:10 scaffolds enriched with in vitro-differentiated neural cells for repairing rat sciatic nerve defects. Tissue Engineering Part A. 2008;14(6):979–93.

[49] Luis AL, Rodrigues JM, Geuna S, Amado S, Simoes MJ, Fregnan F, et al. Neural cell transplantation effects on sciatic nerve regeneration after a standardized crush injury in the rat. Microsurgery. 2008;28(6):458–70.

[50] Pereira T, Gartner A, Amorim I, Almeida A, Caseiro AR, Armada-da-Silva PA, et al. Promoting nerve regeneration in a neurotmesis rat model using poly(DL-lactide-epsilon-caprolactone) membranes and mesenchymal stem cells from the Wharton's jelly: in vitro and in vivo analysis. Biomed Research International. 2014;2014:302659.

[51] Pereira T, Gärtner A, Amorim I, Ribeiro J, França M, Armada-da-Silva P, et al., editors. Development of a skeletal muscle injury model in the rat and in vivo evaluation of the use of human mesenchymal stem cells (HMSCs) from the umbilical cord matrix in myectomy injury treatment. 17th annual Congress of the European College of Sport Science; July 4–7, 2012; Bruges, Belgium, 2012.

[52] Pereira T, Gärtner A, Amorim I, Ribeiro J, França M, Armada-da-Silva P, et al. In vivo evaluation of muscular regeneration treated with human mesenchymal stem cells (HMSCs) from the umbilical cord matrix and different biomaterials in a rat anterior tibial myectomy model. Journal of Tissue Engineering and Regenerative Medicine. 2012;6(Suppl. 1):396.

[53] Ribeiro J, Pereira T, Amorim I, Caseiro AR, Lopes MA, Lima J, et al. Cell therapy with human MSCs isolated from the umbilical cord Wharton jelly associated to a PVA membrane in the treatment of chronic skin wounds. International Journal of Medical Sciences. 2014;11(10):979–87.

[54] Pereira T, Armada-da Silva P, Amorim I, Rêma A, Caseiro A, Gärtner A, et al. Effects of human mesenchymal stem cells isolated from Wharton's jelly of the umbilical cord and conditioned media on skeletal muscle regeneration using a myectomy model. Stem Cells International. 2014;2014.

[55] Friedenstein A, Chailakhjan R, Lalykina K. The development of fibroblast colonies in monolayer cultures of guinea-pig bone marrow and spleen cells. Cell Proliferation. 1970;3(4):393–403.

[56] Lajtha LG. Stem cells and their properties. Proceedings Canadian Cancer Conference. 1967;7:31–9.

[57] Friedenstein AJ, Petrakova KV, Kurolesova AI, Frolova GP. Heterotopic of bone marrow. Analysis of precursor cells for osteogenic and hematopoietic tissues. Transplantation. 1968;6(2):230–47.

[58] Chamberlain G, Fox J, Ashton B, Middleton J. Concise review: mesenchymal stem cells: their phenotype, differentiation capacity, immunological features, and potential for homing. Stem Cells. 2007;25(11):2739–49.

[59] Triffitt JT. Stem cells and the philosopher's stone. Journal of Cellular Biochemistry. Supplement. 2002;38:13–9.

[60] da Silva Meirelles L, Chagastelles PC, Nardi NB. Mesenchymal stem cells reside in virtually all post-natal organs and tissues. Journal of Cell Science. 2006;119(11):2204–13.

[61] Muguruma Y, Reyes M, Nakamura Y, Sato T, Matsuzawa H, Miyatake H, et al. In vivo and in vitro differentiation of myocytes from human bone marrow-derived multipotent progenitor cells. Experimental Hematology. 2003;31(12):1323–30.

[62] Dominici M, Le Blanc K, Mueller I, Slaper-Cortenbach I, Marini F, Krause D, et al. Minimal criteria for defining multipotent mesenchymal stromal cells. The International Society for Cellular Therapy position statement. Cytotherapy. 2006;8(4):315–7.

[63] Caplan AI. Why are MSCs therapeutic? New data: new insight. Journal of Pathology. 2009;217(2):318–24.

[64] Le Blanc K, Ringden O. Immunobiology of human mesenchymal stem cells and future use in hematopoietic stem cell transplantation. Biology of blood and marrow transplantation. Journal of the American Society for Blood and Marrow Transplantation. 2005;11(5):321–34.

[65] Fossett E, Khan WS. Optimising human mesenchymal stem cell numbers for clinical application: a literature review. Stem Cells International. 2012;2012:465259.

[66] Kim DW, Staples M, Shinozuka K, Pantcheva P, Kang SD, Borlongan CV. Wharton's jelly-derived mesenchymal stem cells: phenotypic characterization and optimizing

their therapeutic potential for clinical applications. International Journal of Molecular Sciences. 2013;14(6):11692–712.

[67] Conconi MT, Burra P, Di Liddo R, Calore C, Turetta M, Bellini S, et al. CD105(+) cells from Wharton's jelly show in vitro and in vivo myogenic differentiative potential. International Journal of Molecular Medicine. 2006;18(6):1089–96.

[68] Wu J, Rojas M, Weiss D, Gross G, Häupl T, Almeida-Porada G, et al. Stem cell-dependent therapies mesenchymal stem cells in chronic inflammatory disorders. Berlin: De Gruyter; 2013.

[69] Kawashima N. Characterisation of dental pulp stem cells: a new horizon for tissue regeneration? Archives of Oral Biology. 2012;57(11):1439–58.

[70] Song M, Kim H, Choi Y, Kim K, Chung C. Skeletal myogenic differentiation of human periodontal ligament stromal cells isolated from orthodontically extracted premolars. Korean Journal of Orthodontics. 2012;42(5):249–54.

[71] Hodgetts S, Stagg K, Sturm M, Edel M, Blancafort P. Long live the stem cell: the use of stem cells isolated from post mortem tissues for translational strategies. International Journal of Biochemistry and Cell Biology. 2014;56C:74–81.

[72] Cheng PH, Snyder B, Fillos D, Ibegbu CC, Huang AH, Chan AW. Postnatal stem/progenitor cells derived from the dental pulp of adult chimpanzee. BMC Cell Biology. 2008;9:20.

[73] De Bari C, Dell'Accio F, Tylzanowski P, Luyten FP. Multipotent mesenchymal stem cells from adult human synovial membrane. Arthritis and Rheumatism. 2001;44(8):1928–42.

[74] de la Garza-Rodea AS, van der Velde-van Dijke I, Boersma H, Gonçalves MA, van Bekkum DW, de Vries AA, et al. Myogenic properties of human mesenchymal stem cells derived from three different sources. Cell Transplantation. 2012;21(1):153–73.

[75] Spath L, Rotilio V, Alessandrini M, Gambara G, De Angelis L, Mancini M, et al. Explant-derived human dental pulp stem cells enhance differentiation and proliferation potentials. Journal of Cellular and Molecular Medicine. 2010;14(6B):1635–44.

[76] Grabowska I, Brzoska E, Gawrysiak A, Streminska W, Moraczewski J, Polanski Z, et al. Restricted myogenic potential of mesenchymal stromal cells isolatedfrom umbilical cord. Cell Transplantation. 2012;21(8):1711–26.

[77] Grabowska I, Streminska W, Janczyk-Ilach K, Machaj EK, Pojda Z, Hoser G, et al. Myogenic potential of mesenchymal stem cells—the case of adhesive fraction of human umbilical cord blood cells. Current Stem Cell Research and Therapy. 2013;8(1):82–90.

[78] Santa María L, Rojas CV, Minguell JJ. Signals from damaged but not undamaged skeletal muscle induce myogenic differentiation of rat bone-marrow-derived mesenchymal stem cells. Experimental Cell Research. 2004;300(2):418–26.

[79] Shi D, Reinecke H, Murry CE, Torok-Storb B. Myogenic fusion of human bone marrow stromal cells, but not hematopoietic cells. Blood. 2004;104(1):290–4.

[80] Bajek A, Drewa T, Joachimiak R, Spoz Z, Gagat M, Bodnar M, et al. Myogenic differentiation of mesenchymal stem cells is induced by striated muscle influences in vitro. Current Signal Transduction Therapy. 2012;7(3):220–7.

[81] Meligy FY, Shigemura K, Behnsawy HM, Fujisawa M, Kawabata M, Shirakawa T. The efficiency of in vitro isolation and myogenic differentiation of MSCs derived from adipose connective tissue, bone marrow, and skeletal muscle tissue. In Vitro Cellular and Developmental Biology Animal. 2012;48(4):203–15.

[82] Gang E, Jeong J, Hong S, Hwang S, Kim S, Yang I, et al. Skeletal myogenic differentiation of mesenchymal stem cells isolated from human umbilical cord blood. Stem Cells. 2004;22:617–24.

[83] Hwang JH, Kim IG, Piao S, Jung AR, Lee JY, Park KD, et al. Combination therapy of human adipose-derived stem cells and basic fibroblast growth factor hydrogel in muscle regeneration. Biomaterials. 2013;34(25):6037–45.

[84] Winkler T, von Roth P, Radojewski P, Urbanski A, Hahn S, Preininger B, et al. Immediate and delayed transplantation of mesenchymal stem cells improve muscle force after skeletal muscle injury in rats. Journal of Tissue Engineering and Regenerative Medicine. 2012;6(Suppl 3):s60–7.

[85] Winkler T, von Roth P, Schuman MR, Sieland K, Stoltenburg-Didinger G, Taupitz M, et al. In vivo visualization of locally transplanted mesenchymal stem cells in the severely injured muscle in rats. Tissue Engineering Part A. 2008;14(7):1149–60.

[86] Von Roth P, Radojewski P, Matziolis G, Duda G, Perka C, Winkler T. Systemic mesenchymal stem cell transplantation for the treatment of skeletal muscle trauma. Journal of Bone and Joint Surgery, British Volume. 2011;93-B(II):134.

[87] Peçanha R, Bagno LL, Ribeiro MB, Robottom Ferreira AB, Moraes MO, Zapata-Sudo G, et al. Adipose-derived stem-cell treatment of skeletal muscle injury. Journal of Bone and Joint Surgery, American Volume. 2012;94(7):609–17.

[88] de la Garza-Rodea AS, van der Velde I, Boersma H, Gonçalves MA, van Bekkum DW, de Vries AA, et al. Long-term contribution of human bone marrow mesenchymal stromal cells to skeletal muscle regeneration in mice. Cell Transplantation. 2011;20(2):217–31.

[89] Natsu K, Ochi M, Mochizuki Y, Hachisuka H, Yanada S, Yasunaga Y. Allogeneic bone marrow-derived mesenchymal stromal cells promote the regeneration of in-

jured skeletal muscle without differentiation into myofibers. Tissue Engineering. 2004;10(7–8):1093–112.

[90] von Roth P, Duda GN, Radojewski P, Preininger B, Perka C, Winkler T. Mesenchymal stem cell therapy following muscle trauma leads to improved muscular regeneration in both male and female rats. Gender Medicine. 2012;9(2):129–36.

[91] Matziolis G, Winkler T, Schaser K, Wiemann M, Krocker D, Tuischer J, et al. Autologous bone marrow-derived cells enhance muscle strength following skeletal muscle crush injury in rats. Tissue Engineering. 2006;12(2):361–7.

[92] Cittadella Vigodarzere G, Mantero S. Skeletal muscle tissue engineering: strategies for volumetric constructs. Frontiers in Physiology. 2014;5:362.

[93] Wolf MT, Dearth CL, Sonnenberg SB, Loboa EG, Badylak SF. Naturally derived and synthetic scaffolds for skeletal muscle reconstruction. Advanced Drug Delivery Reviews. 2014.

[94] Sicari BM, Rubin JP, Dearth CL, Wolf MT, Ambrosio F, Boninger M, et al. An acellular biologic scaffold promotes skeletal muscle formation in mice and humans with volumetric muscle loss. Science Translational Medicine. 2014;6(234):234ra58.

[95] Turner NJ, Yates Jr AJ, Weber DJ, Qureshi IR, Stolz DB, Gilbert TW, et al. Xenogeneic extracellular matrix as an inductive scaffold for regeneration of a functioning musculotendinous junction. Tissue Engineering Part A. 2010;16(11):3309–17.

[96] Gentile NE, Stearns KM, Brown EH, Rubin JP, Boninger ML, Dearth CL, et al. Targeted rehabilitation after extracellular matrix scaffold transplantation for the treatment of volumetric muscle loss. American Journal of Physical Medicine and Rehabilitation. 2014;93(11):S79-S87.

[97] Machingal MA, Corona BT, Walters TJ, Kesireddy V, Koval CN, Dannahower A, et al. A tissue-engineered muscle repair construct for functional restoration of an irrecoverable muscle injury in a murine model. Tissue Engineering Part A. 2011;17(17–18):2291–303.

[98] Merritt EK, Cannon MV, Hammers DW, Le LN, Gokhale R, Sarathy A, et al. Repair of traumatic skeletal muscle injury with bone-marrow-derived mesenchymal stem cells seeded on extracellular matrix. Tissue Engineering Part A. 2010;16(9):2871–81.

[99] Klumpp D, Horch RE, Kneser U, Beier JP. Engineering skeletal muscle tissue—new perspectives in vitro and in vivo. Journal of Cellular and Molecular Medicine. 2010;14(11):2622–9.

[100] Levenberg S, Rouwkema J, Macdonald M, Garfein ES, Kohane DS, Darland DC, et al. Engineering vascularized skeletal muscle tissue. Nature Biotechnology. 2005;23(7): 879-84.

[101] Koffler J, Kaufman-Francis K, Shandalov Y, Egozi D, Amiad Pavlov D, Landesberg A, et al. Improved vascular organization enhances functional integration of engi-

neered skeletal muscle grafts. Proceedings of the National Academy of Sciences. 2011;108(36):14789–94.

[102] Shandalov Y, Egozi D, Koffler J, Dado-Rosenfeld D, Ben-Shimol D, Freiman A, et al. An engineered muscle flap for reconstruction of large soft tissue defects. Proceedings of the National Academy of Sciences. 2014;111(16):6010–5.

[103] Winkler T, von Roth P, Matziolis G, Mehta M, Perka C, Duda GN. Dose–response relationship of mesenchymal stem cell transplantation and functional regeneration after severe skeletal muscle injury in rats. Tissue Engineering Part A. 2009;15(3):487–92.

[104] Pereira T, Gärtner A, Amorim I, Armada-da-Silva P, Gomes R, C. P, et al. Biomaterials and stem cell therapies for injuries associated to skeletal muscular tissues. In: Pignatello R, editor. Advances in Biomaterials Science and Biomedical Applications: InTech; 2013.

[105] Anderson JE, Mitchell CM, McGeachie JK, Grounds MD. The time course of basic fibroblast growth factor expression in crush-injured skeletal muscles of SJL/J and BALB/c mice. Experimental Cell Research. 1995;216(2):325–34.

[106] Chen L, Tredget EE, Wu PYG, Wu Y. Paracrine factors of mesenchymal stem cells recruit macrophages and endothelial lineage cells and enhance wound healing. PloS One. 2008;3(4):e1886.

[107] Pereira T, Ivanova G, Caseiro AR, Barbosa P, Bartolo PJ, Santos JD, et al. MSCs conditioned media and umbilical cord blood plasma metabolomics and composition. PloS One. 2014;9(11):e113769.

[108] Ten Broek RW, Grefte S, Von den Hoff JW. Regulatory factors and cell populations involved in skeletal muscle regeneration. Journal of Cellular Physiology. 2010;224(1): 7–16.

[109] Husmann I, Soulet L, Gautron J, Martelly I, Barritault D. Growth factors in skeletal muscle regeneration. Cytokine and Growth Factor Reviews. 1996;7(3):249–58.

[110] Tidball JG, Dorshkind K, Wehling-Henricks M. Shared signaling systems in myeloid cell-mediated muscle regeneration. Development. 2014;141(6):1184–96.

[111] Villalta SA, Nguyen HX, Deng B, Gotoh T, Tidball JG. Shifts in macrophage phenotypes and macrophage competition for arginine metabolism affect the severity of muscle pathology in muscular dystrophy. Human Molecular Genetics. 2009;18(3): 482–96.

[112] Heredia JE, Mukundan L, Chen FM, Mueller AA, Deo RC, Locksley RM, et al. Type 2 innate signals stimulate fibro/adipogenic progenitors to facilitate muscle regeneration. Cell. 2013;153(2):376–88.

[113] Hayashi S, Aso H, Watanabe K, Nara H, Rose M, Ohwada S, et al. Sequence of IGF-I, IGF-II, and HGF expression in regenerating skeletal muscle. Histochemistry and Cell Biology. 2004;122(5):427–34.

[114] DO MKQ, Suzuki T, Gerelt B, Sato Y, Mizunoya W, Nakamura M, et al. Time-coordinated prevalence of extracellular HGF, FGF2 and TGF-β3 in crush-injured skeletal muscle. Animal Science Journal. 2012;83(10):712–7.

[115] Tatsumi R, Anderson JE, Nevoret CJ, Halevy O, Allen RE. HGF/SF is present in normal adult skeletal muscle and is capable of activating satellite cells. Developmental Biology. 1998;194(1):114–28.

[116] Miller KJ, Thaloor D, Matteson S, Pavlath GK. Hepatocyte growth factor affects satellite cell activation and differentiation in regenerating skeletal muscle. American Journal of Physiology—Cell Physiology. 2000;278(1):C174–81.

[117] Yamada M, Tatsumi R, Yamanouchi K, Hosoyama T, Shiratsuchi S-i, Sato A, et al. High concentrations of HGF inhibit skeletal muscle satellite cell proliferation in vitro by inducing expression of myostatin: a possible mechanism for reestablishing satellite cell quiescence in vivo. American Journal of Physiology—Cell Physiology. 2010;298(3):C465-C76.

[118] Tatsumi R, Liu X, Pulido A, Morales M, Sakata T, Dial S, et al. Satellite cell activation in stretched skeletal muscle and the role of nitric oxide and hepatocyte growth factor. American Journal of Physiology—Cell Physiology. 2006;290(6):C1487-C94.

[119] Suzuki S, Yamanouchi K, Soeta C, Katakai Y, Harada R, Naito K, et al. Skeletal muscle injury induces hepatocyte growth factor expression in spleen. Biochemical and Biophysical Research Communications. 2002;292(3):709–14.

[120] Menetrey J, Kasemkijwattana C, Day C, Bosch P, Vogt M, Fu F, et al. Growth factors improve muscle healing in vivo. Journal of Bone and Joint Surgery, British Volume. 2000;82(1):131.

[121] Zanou N, Gailly P. Skeletal muscle hypertrophy and regeneration: interplay between the myogenic regulatory factors (MRFs) and insulin-like growth factors (IGFs) pathways. Cellular and Molecular Life Sciences. 2013;70(21):4117–30.

[122] Pelosi L, Giacinti C, Nardis C, Borsellino G, Rizzuto E, Nicoletti C, et al. Local expression of IGF-1 accelerates muscle regeneration by rapidly modulating inflammatory cytokines and chemokines. FASEB Journal. 2007;21(7):1393–402.

[123] Borselli C, Storrie H, Benesch-Lee F, Shvartsman D, Cezar C, Lichtman JW, et al. Functional muscle regeneration with combined delivery of angiogenesis and myogenesis factors. Proceedings of the National Academy of Sciences. 2010;107(8):3287–92.

[124] Floss T, Arnold H-H, Braun T. A role for FGF-6 in skeletal muscle regeneration. Genes and Development. 1997;11(16):2040–51.

[125] Armand AS, Pariset C, Laziz I, Launay T, Fiore F, Della Gaspera B, et al. FGF6 regulates muscle differentiation through a calcineurin-dependent pathway in regenerating soleus of adult mice. Journal of Cellular Physiology. 2005;204(1):297–308.

[126] Armand A-S, Launay T, Pariset C, Della Gaspera B, Charbonnier F, Chanoine C. Injection of FGF6 accelerates regeneration of the soleus muscle in adult mice. Biochimica et Biophysica Acta (BBA)—Molecular Cell Research. 2003;1642(1–2):97–105.

[127] Messina S, Mazzeo A, Bitto A, Aguennouz Mh, Migliorato A, De Pasquale MG, et al. VEGF overexpression via adeno-associated virus gene transfer promotes skeletal muscle regeneration and enhances muscle function in mdx mice. FASEB Journal. 2007;21(13):3737–46.

[128] Frey SP, Jansen H, Raschke MJ, Meffert RH, Ochman S. VEGF improves skeletal muscle regeneration after acute trauma and reconstruction of the limb in a rabbit model. Clinical Orthopaedics and Related Research. 2012;470(12):3607–14.

[129] Karvinen H, Pasanen E, Rissanen T, Korpisalo P, Vähäkangas E, Jazwa A, et al. Long-term VEGF-A expression promotes aberrant angiogenesis and fibrosis in skeletal muscle. Gene Therapy. 2011;18(12):1166–72.

[130] Hara M, Yuasa S, Shimoji K, Onizuka T, Hayashiji N, Ohno Y, et al. G-CSF influences mouse skeletal muscle development and regeneration by stimulating myoblast proliferation. Journal of Experimental Medicine. 2011;208(4):715–27.

[131] Stratos I, Rotter R, Eipel C, Mittlmeier T, Vollmar B. Granulocyte-colony stimulating factor enhances muscle proliferation and strength following skeletal muscle injury in rats. Journal of Applied Physiology. 2007;103(5):1857–63.

[132] Li Y, Foster W, Deasy BM, Chan Y, Prisk V, Tang Y, et al. Transforming growth factor-β1 induces the differentiation of myogenic cells into fibrotic cells in injured skeletal muscle: a key event in muscle fibrogenesis. American Journal of Pathology. 2004;164(3):1007–19.

[133] Hwang JH, Ra Y-J, Lee KM, Lee JY, Ghil SH. Therapeutic effect of passive mobilization exercise on improvement of muscle regeneration and prevention of fibrosis after laceration injury of rat. Archives of Physical Medicine and Rehabilitation. 2006;87(1): 20–6.

[134] Baraniak PR, McDevitt TC. Stem cell paracrine actions and tissue regeneration. Regenerative Medicine. 2010;5(1):121–43.

[135] Haynesworth SE, Baber MA, Caplan AI. Cytokine expression by human marrow-derived mesenchymal progenitor cells in vitro: effects of dexamethasone and IL-1α. Journal of Cellular Physiology. 1996;166(3):585–92.

[136] Chen JS, Wong VW, Gurtner GC. Therapeutic potential of bone marrow-derived mesenchymal stem cells in cutaneous wound healing. Frontiers in Immunology. 2012;3.

[137] Rehman J, Traktuev D, Li J, Merfeld-Clauss S, Temm-Grove CJ, Bovenkerk JE, et al. Secretion of angiogenic and antiapoptotic factors by human adipose stromal cells. Circulation. 2004;109(10):1292–8.

[138] da Silva Meirelles L, Fontes AM, Covas DT, Caplan AI. Mechanisms involved in the therapeutic properties of mesenchymal stem cells. Cytokine and Growth Factor Reviews. 2009;20(5–6):419–27.

[139] Doorn J, Moll G, Le Blanc K, van Blitterswijk C, de Boer J. Therapeutic applications of mesenchymal stromal cells: paracrine effects and potential improvements. Tissue Engineering Part B: Reviews. 2012;18(2):101–15.

[140] J Salgado A, L Reis R, Sousa N, M Gimble J. Adipose tissue derived stem cells secretome: soluble factors and their roles in regenerative medicine. Current Stem Cell Research and Therapy. 2010;5(2):103–10.

[141] Carvalho MM, Teixeira FG, Reis RL, Sousa N, Salgado AJ. Mesenchymal stem cells in the umbilical cord: phenotypic characterization, secretome and applications in central nervous system regenerative medicine. Current Stem Cell Research and Therapy. 2011;6(3):221–8.

[142] Yoo KH, Jang IK, Lee MW, Kim HE, Yang MS, Eom Y, et al. Comparison of immunomodulatory properties of mesenchymal stem cells derived from adult human tissues. Cellular Immunology. 2009;259(2):150–6.

[143] Teixeira F, Pires A, Fraga J, Carvalho M, Silva N, Panchalingam K, et al. Secretome of mesenchymal stem cells as a new tool for central nervous system regenerative medicine. Journal of Tissue Engineering and Regenerative Medicine. 2014;8:298–9.

[144] Teixeira FG, Carvalho MM, Sousa N, Salgado AJ. Mesenchymal stem cells secretome: a new paradigm for central nervous system regeneration? Cellular and Molecular Life Sciences. 2013;70(20):3871–82.

Chapter 9

Mesenchymal Stem Cells — Their Antimicrobial Effects and Their Promising Future Role as Novel Therapies of Infectious Complications in High Risk Patients

K.A. Al-Anazi and A.M. Al-Jasser

Additional information is available at the end of the chapter

http://dx.doi.org/10.5772/60640

Abstract

Mesenchymal stem cells (MSCs) are heterogeneous progenitor cells that have the capacity of self-renewal and multi-lineage differentiation. These adult stem cells can be derived from several sources including bone marrow (BM), peripheral blood, cord blood, placenta, amniotic fluid, skin and adipose tissue. They have certain distinguishing features and their immunomodulatory and immunosuppressive properties enable them to have several therapeutic and clinical applications. Recently, MSCs have gained enormous potential as they can potentially cure various intractable and chronic diseases and as they have shown effectiveness in the treatment of various infections in animal models and in early clinical trials. MSCs are essential constituents of the framework that supports organ integrity and tissue barriers. Suppression of both T and B cells allows them to be major players in the innate response to bacterial infection and in controlling inflammatory response. Human BM-MSCs possess direct antibacterial activity against Gram-negative bacilli and they have been shown to improve survival and reduce mortality in animal models having septic complications. BM-MSCs are effective in treating sepsis and acute respiratory distress syndrome in high-risk patients such as those with malignant hematological disorders, recipients of solid organ and hematopoietic stem cell transplantation (HSCT) and patients receiving advanced level of care in intensive care units. Additionally, human BM-MSCs can act as drug delivery vehicles by enhancing the effectiveness of conventional antimicrobials and thus they may prevent the evolution of drug-resistant microbes. MSCs contain a subset of interleukin-17+ that is capable of inhibiting the growth of

INTECH
open science | open minds

© 2015 The Author(s). Licensee InTech. This chapter is distributed under the terms of the Creative Commons Attribution License (http://creativecommons.org/licenses/by/3.0), which permits unrestricted use, distribution, and reproduction in any medium, provided the original work is properly cited.

Candida albicans (*C. albicans*). Also, CD 271+ BM-MSCs may provide a long-term protective intracellular niche in the host where *Mycobacterium tuberculosis* (M.TB) organisms remain viable but in a dormant state. Two recent clinical trials in humans that included 57 patients have shown that autologous transplantation of MSCs can successfully treat multidrug resistant (MDR) strains of M.TB. Animal studies have demonstrated that MSCs enhance host defenses against malaria. MSC therapy improves liver function and promotes hepatocellular regeneration in patients with hepatic fibrosis caused by schistosomiasis. Transplantation of MSCs has been shown to reverse right ventricular dilatation, cardiomyopathy and advanced cardiac involvement caused by *Trypanosoma cruzi* infection.

Autologous MSC transfusion in patients having liver cirrhosis secondary to hepatitis B or C infection improves liver function tests. Transfusion of MSCs can confer resistance to human *immunodeficiency virus (HIV)*and may restore immune reconstitution in infected individuals. Also, MSCs obtained from Wharton's jelly of the umbilical cord may become a novel therapy to reverse immune deficiency in individuals infected with HIV_1, particularly immune non-responders. Additionally, recent studies have demonstrated that hematopoietic stem cell-based gene therapy may ultimately offer a curative therapeutic option for *HIV* disease. MSCs improve murine models of acute myocarditis induced by infection with *Coxsackie* B_3 *virus.*

There is low risk of transmission of human herpes viruses by transplantation of MSCs from healthy seropositive donors. Cytomegalovirus (*CMV*) infection impairs the immunosuppressive and antimicrobial effector functions of human MSCs, thus overt *CMV* infection in recipients of HSCT may undermine the clinical efficacy of MSCs in treating graft versus host disease (GVHD). The therapeutic applications of BM-MSCs in recipients of HSCT include: prevention and treatment of GVHD, induction of faster engraftment, immune reconstitution, healing of inflammation as well as prevention and treatment of various infectious complications.

Thus, taking into consideration the remarkable success in the utilization of MSCs in the treatment of various infections in animal models and in human clinical trials, it is reasonable to predict that MSCS may become very promising novel therapeutic modalities as they have the potential to control or even cure various infectious complications in high-risk patients. However, the field is still in its infancy and plenty of research and clinical trials are required to refine their therapeutic indications. Banking of MSCs is vital to make them available for use. Finally; strict guidelines, standardization techniques and quality control measures are urgently required for collection, cryopreservation and clinical utilization of MSCs.

Keywords: Mesenchymal stem cells, Bone marrow, Wharton's jelly, Sepsis, Multi-drug resistance

http://dx.doi.org/10.5772/60640

1. Introduction

Mesenchymal stem cells (MSCs) - also called fibroblastoid cells, giant fat cells with blanket cells, spindle-shaped flattened cells or very small round cells - have an inherent ability of self-renewal, proliferation and differentiation toward mature tissues depending on the surrounding microenvironment [1, 2]. Such characteristic, intrinsic to stem cells, makes MSC very attractive for utilization in cell therapy and regenerative medicine [2]. The discovery of MSCs can be dated to the 1960s and can be credited to the work of AJ Friedenstein during which he observed that bone marrow (BM) is the source of MSCs in postnatal life [1].

MSCs comprise only a tiny fraction of BM and other tissues. BM-derived MSCs [BM-MSCs] constitute 0.0001% to 0.01% of all nucleated BM cells [2]. Adipose tissues contain 100, 000 MSCs in each gram of fat. Adipose tissue-derived MSCs (AT-MSCs) represent an attractive cell source for stem cell therapy. Transplantation of up to $2x10^8$ cells / kg of autologous human adipose tissue-derived MSCs may be safe when given by slow intravenous infusion [2]. The sources of MSCs, their distinguishing features and their therapeutic indications are included in Table 1 [1-4].

Sources of MSCs	Distinguishing features of MSCs	Therapeutic indications of MSCs
1- Bone marrow	(A) They must be plastic adherent	1- In stem cell transplantation:
2- Peripheral blood	(B) They must be capable of	- Prevention and treatment of graft versus
3- Blood of umbilical cord	differentiating into:	host disease
4- Placenta	osteoblasts, adipocytes	- Enhancement of engraftment
4- Amniotic fluid	and chondrocytes	2- In cardiology; treatment of:
5- Synovial fluid	(C) Findings on flowcytometry:	- Coronary artery disease
6- Dental tissues e.g. pulp	[1] positive surface molecules:	- Myocardial infarction
7- Palatal tonsil	- CD 105	- Dilated cardiomyopathy
8- Fallopian tubes	- CD 73	3- Critical limb ischemia
9- Fat: adipose tissue	- CD 90	4- Repair of skeletal tissues
10- Parathyroid glands	[2] negative surface molecules:	5- Non-healing chronic wounds
	- CD 45	6- Chronic spinal cord injuries
	- CD 34	7- Liver injury
	- CD 14	8- Regenerative medicine
	- CD 11	9- Osteogenesis imperfecta
	- CD 19	10- Amyotrophic lateral sclerosis
	- CD 79	11- Acute respiratory distress syndrome
	- HLA - DR	12- Severe autoimmune disorders: systemic
		lupus erythromatosis, rheumatoid arthritis,
		multiple sclerosis and systemic sclerosis

MSCs: mesenchymal stem cells

Table 1. Sources, basic features and therapeutic indications of MSCs

Study and year of publication	Type of MSCs, route of administration of MSCs and sepsis model	Results and proposed mechanism(s) of action
Hau SR et al 2013	- MSCs harvested from compact bone of mice - Intravenous injection of MSCs into tail veins of mice - CLP, in vivo method - Polymicrobial sepsis	- Increased survival of mice - Decreased organ injury - Increased neutrophil phagocytosis - Decreased circulating bacteria
Mei SHJ et al 2010	- Bone Marrow-MSCs - Intravenous administration - CLP - In vivo method	- Significant reduction in mortality in septic mice receiving appropriate anti-microbial therapy. - Significant increase in bacterial clearance partly due to increased phagocytic activity of host immune cells
Luo CT et al 2014	- Bone marrow-MSCs - Intravenous administration - CLP - Polymicrobial sepsis	- Decreased circulating bacteria - Alleviation of sepsis - related acute kidney injury - Improved survival of mice. - Inhibition of IL-17 secretion - After MSC therapy, the following interleukins and inflammatory mediators were reduced in kidney tissues: Il-6, IL-7, TNF-α, INF-γ, $CXCL_1$, CXCL2, CXCL5, CCL2 and CCL3
Gupta N et al 2012	- Bone marrow - MSCs - Intratracheal injection - Unstimulated mouse MSCs - *E. Coli* pneumonia - In vitro study	- Increased survival of mice - Inhibition of bacterial growth - Increased lipocalin-2 in bronchoalvealar lavage fluid. - Increased phagocytic activity
Nemeth K et al 2009	- Bone marrow - MSCs - Activated MSCs - CLP - In vivo study	- Reduced mortality of mice - Improved organ function - The beneficial effects of MSCs were eliminated by: macrophage depletion or pre-treat-IL-10 receptors. - MSCs reprogram macrophages by releasing postaglandin E2.
Gonzalez-Rey et al 2008	- Human and murine adipose tissue-MSCs - Intraperitoneal injection - CLP and endotoxin injection - Colitis and sepsis model, in vivo study	- Systemic infusion of adipose tissue-derived MSCs resulted in: increased survival of mice in addition to significant amelioration of severity of colitis and sepsis.
Chang C-L Et al 2012	- Apoptotic adipose tissue- MSCs - Rat model - Penile venous transfusion	- Protection of major organs from damage - Improved prognosis and reduced mortality of animals treated with MSCs.

Study and year of publication	Type of MSCs, route of administration of MSCs and sepsis model	Results and proposed mechanism(s) of action
	- CLP - Peritoneal sepsis, acute kidney injury	- Increased circulating levels of TNF-γ - Reduction in the numbers of: helper and cytotoxic T-cells, T-regulatory cells in peripheral blood and spleens of animals.
Yang R et al 2013	- IL-17 producing MSCs	- Inhibition of growth of *Candiada albicans* - Mechanism of action: stimulation of IL-17 production.
Sung P-H et al 2013	- Rat model - Adipose tissue - MSCs - CLP - Sepsis syndrome, kidney and lung injury	- Adipose tissue MSC therapy was superior to healthy adipose-derived MSC therapy in preventing major organ damage in rats treated with CLP-induced sepsis syndrome.

MSCs: mesenchymal stem cells; IL: interleukin

TNF: tumor necrosis factor; CLP: cecal ligation puncture

INF: interferon

Table 2. Studies using murine MSCs in infection or sepsis models

Study and year of publication	Type of MSCs, route of administration and sepsis model	Results and possible mechanism(s) of action
Meisel R et al 2011	- Human MSCs - Interferon –γ stimulated MSCs - In vitro study	- Inhibition of gram-positive bacteria - Inhibition of intracellular CMV and HSV-1 replication - Increased indoleamine -2,3- dioxygenase
Krasnodembskaya A et al 2010	- Human BM-MSCs murine model - Intravenous admin istration of MSCs - Intratracheal *E.coli, Pseudomonas aeruginosa* and *Staphylococcus aureus* pneumonia	- Increased antimicrobial activity and decreased bacterial growth in bronchoalveolar lavage. - Decrease in lung bacterial load. - Inhibition of bacterial growth in vitro - IV MSCs + neutralizing antibodies to IL-37: reduced bacterial clearance.
Krasnodembskaya A et al 2012	- Human BM - MSCs - Intravenous administration - Murine Gram-negative peritoneal sepsis	- Increased survival due to increased clearance of bacteria from circulation. -Increased phagocytic activity of mononuclear cells in spleens of mice. - Marked reduction in circulating bacteria in the blood of treated mice. - Increased number of platelets inhibitor -1 (PAI-1) - No change in cytokine levels in plasma and peritoneal fluid.
Kim ES et al	- Human umbilical cord blood - MSCs	- Increased survival of treated mice - Reduced bacterial burden in pneumonia model.

Study and year of publication	Type of MSCs, route of administration and sepsis model	Results and possible mechanism(s) of action
2011	- Intrathecal administration - E. coli induced acute lung injury - In vivo study in mice	- Acute lung injury induced by *E.coli* was attenuated by: down-modulation of inflammatory process and enhancement of bacterial clearance - Anti-inflammatory mechanisms involved.
Casatella MA et al 2011	- Human MSCs - In vitro study	- Enhancement of neutrophil function - Proposed mechanisms: IL- 6, INF- β and GM-CSF
Maqbool M et al 2011	- Human MSCs - In vitro study	- Inhibition of neutrophil apoptosis - Unknown mechanism of action - No elucidation of molecular mechanisms that govern protection of neutrophils from serum-induced apoptosis
Rafaghello L et al 2008	- Human BM-MSCs and neutrophils - In vitro, MSC - conditioned media - Delivery method: co-culture - Effect of BM-MSCs on neutrophil survival and effector function.	- Inhibition of apoptosis of resting and IL-8 activated neutrophils with both MSCs and MSC-conditioned media - Recombinant IL-6 was found to protect neutrophils from apoptosis - Effect was mediated by IL-6
Lee JW et al 2013	- Human BM derived MSCs - Ex-vivo perfused human lung injured and intrabronchial *E. coli* delivery - Bacterial induction of acute lung injury * IV or intrabronchial injection of MSCs	- Increased macrophage phagocytosis - Reduced alveolar bacterial burden - Intrabronchial KGF-7 caused: increased alveolar macrophage phagocytosis and decreased bacterial load. - KGF duplicated most of the antimicrobial effects of MSCs. - In vitro: KGF-7 positive monocytes increased bacterial killing and monocyte survival and FGF-7 blocking antibodies nullified antimicrobial effects ex vivo and in vitro.

MSCs: mesenchymal stem cells; IV: intravenous

IL: interleukin; KGF: keratinocyte growth factor

IFN: interferon; GM-CSF: granulocyte monocyte-colony stimulating factor

Table 3. Studies using human MSCs in infection or sepsis models

Based on current clinical trials and irrespective of the treated disease condition or the mode of administration, MSC therapy appears relatively safe [3, 5]. Therapeutic applications of MSCs raise a series of concerns about the safety of culture-expanded MSCs for human use [4]. However, further large scale clinical trials with rigorous reporting of adverse events are required to further define the safety profile of MSCs [5]. Complications of MSC therapies include: (1) febrile reactions during or shortly after transfusion, (2) tumor modulation and malignant transformation, (3) increased risk of cancer, particularly hematologic malignancy, (4) genetic manipulation, induction of genetic instability and evolution of chromosomal abnormalities, (5) increased incidence of infectious complications, particularly in recipients of hematopoietic stem cell transplantation (HSCT) having graft versus host disease (GVHD) treated with MSCs, and (6) persistence of human *Parvovirus* B_{19} in multipotent MSCs, expressing the erythrocyte P antigen, that are used to improve the outcome of HSCT and in regenerative medicine [1, 6-9]. Therefore, strict control and safety measures are needed in the production of MSCs for cell-based therapies in order to minimize or abolish the risk of malignant transformation [1]. It is also essential to establish standardized manufacture guidelines for the isolation, expansion, preservation and delivery of MSCs for safety evaluation and clinical applications [4].

2. MSCs and sepsis

Sepsis is the leading cause of death in the intensive care units and it ranks among the top 10 causes of death in the general population worldwide. Despite the advances in medical care, mortality rates from sepsis remain as high as 30%-50% in severe cases [10]. Sepsis is a systemic inflammatory response that is typically triggered by bacterial or other microbial infection and it involves a complex interaction between the microbial pathogen and the host immune system [11, 12]. The inflammatory response, designed to control infection, affects end-organ perfusion and exacerbates tissue injury leading to multi-organ failure [11]. In the early phase of sepsis, endogenous proinflammatory cytokines and coagulation pathways become hyperactive and their adverse effects cause multi-organ failure, collapse of the circulatory system and ultimately death of the affected individual [12].

Sepsis is associated with a surge of systemic signaling molecules including cytokines and growth factors that may become uncontrollable [11]. Stem cells have profound effects on the inflammatory response and coagulation cascade that are activated during sepsis [13]. Stem cells have been shown to modulate more than 3000 genes in experimental sepsis models. They enhance the clearance of pathogens and repair of injured tissues in sepsis. In addition, stem cells have been found to have an impact on cell leak that is encountered during sepsis [13].

2.1. MSCs in sepsis and experimental sepsis models

Mesenchymal stroma is an essential component of tissue barriers and it participates in protecting the body from infections. A breach in the integrity of tissue barriers such as trauma can lead to translocation of bacteria and colonization of vital organs, thus ultimately leading to multiple organ failure and sepsis [14]. Tissue injury compromises barrier function and increases the risk of infection and sepsis. Tissue injury can induce recruitment of MSCs from BM and promote their proliferation so as to reconstitute the integrity of injured tissues and mediate natural debridement [14]. MSCs can modulate systemic response to bacterial infection and support tissue repair in addition to healing after being recruited at the sites of tissue injury [14]. Sepsis is a devastating condition characterized by systemic activation of inflammatory and coagulation pathways in response to microbial invasion of sterile body organs or tissues [15]. Severe sepsis, sepsis with at least one dysfunctional organ, is a leading cause of death in intensive care units (ICUs) as it is associated with mortality rates ranging between 30% and 50% [15].

The development of experimental sepsis models to elucidate the pathophysiology and progression of clinical sepsis spans over the last 8 decades. The following transitions in animal models took place since the 1930s: endotoxemia, bacteremia, ischemia and bowel perforation and finally cecal ligation puncture (CLP) and the colon ascends stent peritonitis (CASP) models [15]. Genetic differences in mice and possibly humans are associated with differences in the inflammatory process initiated in response to sepsis and that ultimately affect the outcome of sepsis. Hence, transgenic animal models have been extensively utilized in sepsis research [15]. Upon stimulation by endotoxins or proinflammatory mediators, activated neutrophils release a chromatin material composed of DNA and antimicrobial granular proteins in the form of neutrophil extra-cellular traps (NETs). The formation of NETs, NETOSIS, is an emerging field in sepsis research [15]. However, the presence of NETOSIS is distinct from that of necrosis and apoptosis. NETOSIS can be experimentally induced by endotoxins such as lipopolysaccharides (LPS), Gram-negative as well as Gram-positive bacteria, fungi and proinflammatory cytokines such as activated platelets and interleukin-8 (IL-8). NET formation has been found to cause host tissue and cellular damage [15]. LPS-induced endotoxemia in mice as well as plasma obtained from humans with severe sepsis were able to trigger NETOSIS. In vitro, NET formation resulted in endothelial cell damage while in vivo it caused hepatotoxicity in LPS-challenged mice [15]. In septic hosts, NETs can exacerbate sepsis by releasing high concentrations of potent proteases, forming a chromatin meshwork and trapping host cells (erythrocytes, leukocytes and platelets) that can potentiate inflammation, coagulation and ischemia in involved tissues [15].

Microparticles are a heterogenous population of small membrane-coated vesicles that are released by several cell lines upon activation or apoptosis [16]. Exosomes are a specialized category of microparticles with specific functions in immune response and protein sorting. Exosomes are released mainly from antigen presenting cells, although they have been identified after activation of platelets and mast cells and in body fluids such as urine or bronchoalveolar lavage [16]. Although the role of exosomes in sepsis remains deeply unexplored, accumulating data suggest that microparticles and exosomes play a role in the three pathways clearly involved in the pathogenesis of sepsis: inflammation, thrombosis and vascular dysfunction [16].

Human MSCs possess antimicrobial and immunosuppressive effects that are partly mediated by the tryptophan catalyzing enzyme indoleamine-2, 3-dioxygenase (IDO) [17]. Upon stimulation by inflammatory cytokines, MSCs exhibit broad spectrum antimicrobial effector functions directed against various clinically relevant pathogens and these effects are dependent on IDO and/or the antimicrobial peptide LL-37 [17]. MSCs have antiapoptotic, antifibrotic and angiogenic properties. Interferon γ (IFN-γ) primes MSC-mediated immunoregulatory effects and induces nitrous oxide (NO) production in MSCs. NO exerts antiapoptotic effects on cardiomyocytes and has antiviral properties [18].

The role of microparticles and exosomes in mediating vascular dysfunction suggests that they may represent novel pathways in paracrine transcellular signaling in vascular microenvironment [16]. Further studies may aid in the clarification of their exact effects in sepsis and the development of additional interventional strategies for the prevention and treatment of sepsis [16]. Examples of experimental sepsis and infection models that utilized human as well as murine MSCs are shown in Tables 2 and 3 [19-35].

MSCs function at several levels of the inflammatory response, particularly in the early phase of sepsis, to regulate a wide panel of inflammatory cytokines and inhibit leukocyte infiltration into several target organs [12]. In an endotoxemic rat model, human AT-MSCs had the following beneficial effects: (1) modulation of host responses, (2) reduction of inflammatory cytokine levels in serum and lung, (3) reduction of alveolar inflammatory cell infiltration in the lung, (4) reduction of liver injury, and (5) improvement of tissue hypoperfusion [12]. As systemic administration of human-AT-MSCs at the onset of endotoxemia ameliorated the serological and histological signs of endotoxemia, they may become attractive candidates for cell therapy in the treatment of endotoxemia and septic shock [12].

Increasing evidence suggests that BM-MSCs secrete molecules that inhibit the effector function of immune cells [11]. AT-MSCs are a recently discovered cell population with much in common to their BM-derived counterparts. Both BM-MSCs and AT-MSCs have been shown to be effective in certain pre-clinical studies as they effectively inhibit the activation of T-cells [11].

Cellular immunotherapy for septic shock (CISS) trial is the first trial of stem cell therapy in humans [13]. It has 2 phases and the objectives of this Canadian trial are as follows: (1) in phase I, to establish safety and patient tolerability of stem cells and to find the optimal dose of stem cells to be administered, and (2) in phase II, to look at loose surrogates of efficacy for patients with septic shock [13]. The study has two arms: (1) an observational arm, where patients do not receive stem cells, and (2) an interventional arm, where patients receive stem cells with three different dose schedules [13]. The study is expected to address many aspects of stem cell therapy in patients having septic shock such as safety, tolerable dose and feasibility of use [13, 36].

In animal models of *Staphylococcal* toxic shock syndrome, MSCs have been shown to: (1) enhance bacterial clearance, and (2) suppress proinflammatory cytokine production, but they failed to prolong survival in experimental models [37, 38]. MSC therapy has also been shown to: (1) enhance antibiotic therapy in chronic *Staphylococcal* infections, and (2) enhance the effectiveness of conventional antibiotics in the treatment of antibiotic-resistant microbial infections [39, 40]. In patients with sepsis, the clinical application of MSC-based therapy is feasible, well tolerated and may be beneficial [36]. Studies have also shown that MSCs not only modulate the systemic inflammation, but also improve cardiac function in patients having endotoxemia [41]. Additionally, BM-MSCs can uptake and release antimicrobials, for example ciprofloxacin, into infected deep environment such as chronic osteomyelitis or deep seated abscesses [42]. Also, MSCs can act as drug delivery system for antibiotics or vehicles for targeted therapies in order to enhance effects of antimicrobials and other targeted therapies [42, 43].

2.2. Conclusions that can be drawn from experimental models

MSCs may be a potential new therapeutic modality in the prevention and reduction of sepsis-associated lung injury [21]. Decreased systemic and pulmonary cytokine levels may prevent acute lung injury and organ dysfunction. Genetic effects downregulate inflammation-related genes such as IL-10 and IL-6 and upregulate genes involved in the promotion of phagocytosis and bacterial killing [20]. The bacterial clearance effect could be party due to the upregulation of lipocalin -2 production by MSCs. However, lipocalin-2 antibodies have been found to block antimicrobial effects of MSCs [22].

In peritoneal sepsis models using murine MSCs; the following mechanisms were involved: down regulation of Th_1 inflammatory responses, reduced inflammatory cytokines and chemokines as well as increased IL-10 level and induction of IL-10 T-regulatory cells [24]. Apoptotic AT-MSCs represent an endogenous therapeutic strategy that may be enhanced for maximum clinical benefits [25]. Treatment with apoptotic AT-MSCs caused attenuation of sepsis syndrome induced lung and kidney parenchymal injury through suppression of inflammation, apoptosis and oxidative stress and enhancement of anti-oxidation and anti-apoptosis in rodent models [27].

Human BM-MSCs possess direct antimicrobial activity that is mediated by secretion of human cathelicidin hCAP-18/IL-37 [29]. Human MSCs participate in the innate response against Gram-negative bacteria through the secretion of the antimicrobial peptide IL-37 [29]. MSCs inhibit apoptosis of both resting and activated neutrophils and they increase the viability of resting and activated neutrophils in various culture serum concentrations [33]. MSCs can restore alveolar fluid clearance, reduce inflammation and exert antimicrobial activity party through keratinocyte growth factor (KGF) secretion [35]. Lastly, cultured or banked human BM-MSCs may be effective in treating sepsis in high-risk patients [23].

3. MSCs and viral infections

Fetal membrane derived MSCs [FM-MSCs] are susceptible to *herpes simplex virus-1* (*HSV-1*), *HSV-2*, *varicella zoster virus* and human *cytomegalovirus* (*CMV*) in vitro, while *Epstein-Barr virus* (*EBV*), human *herpes virus-7* (*HHV-7*) and *HHV-8* are capable of entering into FM-MSCs, but only limited gene expression occurs, thus resulting in non-productive infection. Therefore, FM-MSCs should be screened for the presence of herpes viruses before xenotransplantation [44]. Transcription factor nuclear factor (NF-kB) may affect the efficacy of *CMV* infection of MSCs. NF-kB inhibitors exacerbate the infection by the activation of human *CMV* in MSCs thus close attention should be given to human *CMV* infection once an NF-kB inhibitor is prescribed in clinical settings [45].

Human *CMV* is a leading cause of life-threatening complications in immunocompromised individuals, such as those with *acquired immune deficiency virus* (*AIDS*) and recipients of solid organ transplant (SOT) as well as HSCT [46]. Human *CMV* infects a wide range of cell types including fibroblasts, monocytes, granulocytes and BM cells such as myeloid progenitor cells and MSCs. Following primary infection, human *CMV* establishes a lifelong latency in the host [46]. CD34+ progenitor cells have been recognized as the most likely reservoir for latent *CMV* infection in the BM, while CD34+ cells from the blood of healthy human *CMV* carriers have been demonstrated to contain the latent human *CMV*. The interaction between human *CMV* and BM-MSCs is complex and it is possible that BM-MSCs infected with human *CMV* may play a role in the development of human *CMV*-associated pathology [46]. Frequent virus reactivation in BM cells and subsequent productive human *CMV* infection of MSCs may be the underlying cause of various pathological processes [46]. *CMV*-infected MSCs lose their cytokine-induced immunosuppressive capacity and become no longer able to restrict microbial growth [17]. IDO expression is substantially impaired following *CMV* infection of MSCs and the interaction between *CMV* infection of MSCs and IDO expression critically depends upon (1) having an intact virus, (2) the number of infected MSCs, and (3) the viral load [17]. It is recommended that patients scheduled for MSC therapy should undergo thorough evaluation for an active *CMV* infection and receive *CMV*-directed therapy prior to the administration of MSCs as overt *CMV* infection in recipients of MSCs may undermine the clinical efficacy of MSC therapy [17].

MSCs could express receptors that permit their infection by *human immunodeficiency virus* (*HIV*)-1 [47]. Highly active antiretroviral therapy (HAART) significantly decreases mortality and morbidity in patients with *HIV*-1 infection, but immune non-responders (INRs) with full viral suppression still fail to reverse the immune deficiency state [48]. In a pilot prospective controlled clinical trial that had enrolled 13 patients with *HIV*-1 infection, it was found that umbilical cord or more precisely Wharton's jelly derived-MSC transfusions were not only well tolerated, but also found to efficiently improve immune reconstitution in INR patients, suggesting that MSC therapy may be used as a novel therapeutic approach to reverse immune deficiency in INR-*HIV*-1 infected individuals [48]. Another study suggested that *HIV*-1 infected individuals could be cured if the HAART therapy is administered before the virus is able to establish its reservoir in the body [49]. Human *T-lymphotropic virus* (*HTLV*)-1 could infect and replicate in human BM-MSCs possibly by involvement or infiltration of CD4+ T-lymphocytes [50].

Although pre-clinical studies have shown that MSC therapy can induce anti-inflammatory effects and enhance repair of the injured lung and despite the clinical and pathological similarities between acute lung injury and severe influenza infection, it was found that MSC therapy may not be effective for the prevention and/or treatment of acute severe influenza [51]. It is well established that MSCs can be infected by *adenoviruses* and that these viruses are widely used as gene transfer vectors [52]. Recent studies revealed that the cationic polymer polybrene can potentiate *Adenovirus*-mediated transgene delivery into MSCs and should be routinely used as a safe, effective and inexpensive augmenting agent for *Adenovirus*-mediated gene transfer in MSCs [52].

Coxsackie virusB_3 (*CVB*$_3$) causes myocarditis not only by immune mediated mechanisms but also by inducing direct cardiomyocyte injury [18]. MSCs improve murine acute *CVB*$_3$-induced myocarditis via antiapoptotic and immunomodulatory mechanisms that occur in a NO-dependent manner and require priming with IFN-γ [18]. MSCs have the potential to treat acute

CVB_3-induced myocarditis since MSCs (1) cannot be infected with *CVB_3*, (2) have antiapoptotic and antiviral effects, and (3) improve cardiac function in experimental models of acute *CVB_3*-induced myocarditis [18].

Several lines of evidence indicate that human BM-MSCs can maintain *hepatitis B virus* (*HBV*) infection in vitro and support the replication of *HBV*-DNA. Human BM-MSCs may be a useful tool for investigating the *HBV* life-cycle and the mechanism of initial virus-cell infection [53].

4. MSCs and *Mycobacterium tuberculosis*

Persisters are a small fraction of quiescent bacterial cells that survive lethal antibiotics or stresses but are capable of reactivation under certain circumstances. Despite the discovery of persisters more than 70 years ago, the mechanisms of persistence are still poorly understood [54]. A number of pathways and genetic abnormalities have recently been identified and they may explain the mechanisms of persistence. These pathways include: DNA repair or protection, toxin-antitoxin modules, phosphate metabolism, antioxidative defense, efflux and macromolecule degradation [54]. However, more sensitive techniques are required to have a better understanding of the mechanisms of persistence. Once the underlying mechanisms are elucidated, cellular therapies, vaccination and targeted treatments are likely to make significant improvements in the management of persistent infections [54].

Bacterial persistence is the hallmark of tuberculosis (TB). The remarkable success of *Mycobacterium TB* (*M.TB*) as a human pathogen is attributed to its ability to program itself into entering prolonged periods of dormancy thus resulting in a latent TB infection [55]. Thus, *M.TB* can successfully evade the host immune system to establish a persistent infection [56, 57]. *M.TB* suppresses T-lymphocyte responses by recruiting MSCs into the site of infection and these cells can inhibit T-lymphocyte responses further by producing NO [56, 57]. Recruitment of MSCs into the tuberculous granuloma plays a vital role in the pathogenesis of *M.TB* infection as these MSCs infiltrate into the sites of infection and position themselves between the harbored pathogen and the effector T-cells [56, 57]. Targeting MSCs or NO is a feasible strategy to design future therapeutic and preventive interventions against *M.TB* infections [56, 57]. Macrophage apoptosis is a rather essential process in the development and maintenance of immunity as its augments the adaptive immune response through dendritic cell activation and subsequent antigen presentation and also reduces bacterial viability [58]. CD_4+T-cells play a crucial role in host defense against *M.TB* infections [59]. Animal studies have shown that isoniazid (INH) treatment causes dramatic reduction in *M.TB* antigen-specific immune responses and the induction of apoptosis in activated CD_4+ cells thus rendering treated animals vulnerable to TB reactivation and reinfection [59]. The finding that anti-TB treatment may be associated with further immune impairment should be taken into consideration in designing new anti-TB therapies [59].

Studies have shown that (1) *M.TB* may maintain long-term intracellular viability in BM-derived CD_{271}+/CD_{45}-MSC population in vitro, and (2) viable *M.TB* organisms have been detected in CD_{271}+/CD_{45}-MSCs isolated from individuals who had successfully completed months of anti-TB chemotherapy [57, 60, 61]. Thus, CD_{271}+BM-MSCs are capable of providing an antimicrobial protective intracellular niche in the host in which dormant *M.TB* can reside for long periods of time [57, 60, 61].

4.1. The role of autologous MSC transplantation in the treatment of TB

Recently, autologous transplantation of MSCs has successfully been utilized in the treatment of infections caused by multi-drug resistant TB (MDR-TB) and even extensively drug resistant TB (XDR-TB) [57, 62, 63]. In one study, 27 patients in whom previous anti-TB drug therapy had been unsuccessful were included and they received autologous MSC transplantation [62]. Positive clinical responses were obtained in all patients, bacterial discharge from lung was abolished in 20 patients and resolution of tissue damage and lung cavitation were achieved in 11 patients 4 months after MSC infusion [62]. Also, persistent remission of the tuberculous process was obtained in 56% of patients who had follow-up for 2 years post MSC transplantation [62]. In a second phase 1 clinical trial, 30 patients with MDR and XDR-TB received autologous MSC transplantation after 4 weeks of starting anti-TB therapy [63]. Six months post-autologous MSC transplantation (1) no major adverse events were encountered and (2) 70% of patients showed radiological improvement, while 16.7% of patients showed stable radiological appearances [63]. Eighteen months post-MSC transplantation, 53% of the patients were cured while 10% of transplanted patients had evidence of treatment failure [63]. Therefore, combining standard anti-TB chemotherapy with autologous MSCT may become a promising maneuver to enhance the efficacy of treatment in patients with drug-resistant pulmonary TB [57, 62, 63].

5. MSCs and parasites

Plasmodium berghei infection in mice leads to massive recruitment of MSCs in secondary lymphoid organs [64, 65]. Also, infusion of MSCs into naive mice can confer host resistance against malaria by (1) augmentation of IL-12 production, (2) suppression of IL-10 production, (3) inhibition of hemozoin, and (4) dramatic reduction in regulatory T-cells in spleens of animals [64, 65]. *Leishmania* parasites can survive in different tissues and organs for decades even after treatment [66]. Several studies have shown that intracellular parasites can persist and remain viable in fibroblasts in latent, inactive or dormant forms [66, 67]. Also, *Leishmania* parasites could remain viable in inactive forms in AT-MSCs in vitro [66]. Therefore, screening for leishmaniasis is essential in recipients of HSCT living in areas that are endemic for leishmaniasis [66].

MSCs ameliorate liver injury and reduce fibrosis in mouse models of *Schistosoma japonicum* (*S. japonicum*). When combined with praziquantel, the preceding effects are enhanced [68, 69]. Infusion of MSCs in mice infected with *S. mansoni* caused (1) reduction of hepatic fibrosis, (2) amelioration of liver injury by induction of liver regeneration, thus ultimately leading to (3) improvement in liver function tests [70]. IFN-γ activated murine MSCs (1) could not inhibit the growth of a highly virulent strain of *Toxoplasma gondii* (BK), (2) strongly inhibited the growth of a type II strain of *Toxoplasma gondii* (ME4a), and (3) inhibited the growth of *Neospora caninum* [71].

5.1. MSCs in Chagas disease

Chagas disease is caused by infection with *Trypanosoma cruzi* (*T. cruzi*) [72, 73]. Up to 30% of infected individuals develop cardiac symptoms related to a chronic chagasic dilated form of cardiomyopathy (CMP) [72]. Cardiac manifestations of Chagas disease include congestive cardiac failure, arrhythmias, heart block, thromboembolism, stroke and sudden death [73]. Available therapies include treatment of heart failure and arrhythmias, antiparasitic therapy, resynchronization treatment, heart transplantation and stem cell therapies [73].

The use of cell therapies to improve cardiac function has been attempted experimentally for more than two decades [72]. The use of BM-derived cells to treat cardiac diseases gained impulse based on the observation that stromal BM cells could be induced to differentiate into cardiomyocytes in vitro and when transplanted into cryoinjured rat hearts improved myocardial function and promoted angiogenesis [72, 74, 75]. Another significant development was that hematopoietic stem cells (HSCs) obtained from transgenic mice expressing enhanced green fluorescent protein, when transplanted into infarcted hearts of syngenic mice, differentiated into cardiac muscle and vascular cells [72, 76]. Since then, many laboratories reported that HSCs and stromal cells derived from BM improved myocardial function in animal models of both cryoinjured and ischemic heart lesions [72]. In one study, cardiac MSCs were isolated from hearts of green fluorescent protein transgenic mice then injected into left ventricular (LV) walls of mice chronically infected with *T. cruzi* [77]. Results of the study showed (1) cardiac MSCs demonstrated adipogenic, osteogenic and differentiation potentials, (2) histological analysis showed that mice treated with cardiac MSCs had a significant reduction in inflammatory cells but no reduction in fibrotic areas, and (3) molecular studies showed that cell therapy significantly decreased TNF-α expression and increased transforming growth factor-beta in heart samples [77]. The results clearly demonstrated that cardiac MSCs exert a protective effect in cardiac chagasic CMP primarily by immunomodulation [77]. In another model of chagasic CMP, direct LV injection of co-cultured skeletal myoblasts and stromal BM-derived cells improved heart function in chronically infected chagasic rats as measured by echocardiography [72, 78]. Injection of the co-cultured cells increased ejection fraction and decreased diastolic and end-systolic volumes [72, 78]. Intraperitoneal administration of MSCs into a mouse model of chronic chagasic CMP reduced inflammation and fibrosis in hearts of mice but had no effect on cardiac function as shown by another study involving an experimental animal model [79].

After the encouraging results in animal models, a clinical trial examining the feasibility and safety of intracoronary injection of autologous BM cell transplantation in patients with congestive cardiac failure due to chronic chagasic CMP was performed [72, 80]. Results of the trial showed that the procedure was safe and effective as cell therapy induced small but rather significant improvements in both ejection fraction and quality of life in patients included in the study [72, 80]. In an experimental cardiac ischemia model, the administration of granulocyte-colony stimulating factor (G-CSF) had beneficial effects on cardiac structure and function

[81]. Repeated administration of G-CSF in another experimental model of chronic chagasic CMP, which closely resembled human disease, induced beneficial effects on (1) cardiac structure as inflammation and fibrosis were reduced, and (2) cardiac function [81].

Based on the promising results of the previous safety trial, a larger multicenter, randomized, double-blinded and placebo-controlled trial was designed to test the efficacy of intra-coronary injection of BM-derived mononuclear cells (MNCs) in patients with chronic chagasic CMP [72, 82]. Although no serious adverse events were observed, this efficacy trial showed no additional benefit of intracoronary injection of BM- MNCs in chagasic patients with low ejection fraction, that is, intracoronary injection of BM-MNCs neither improved LV function nor improved quality of life in patient with chronic chagasic CMP [72, 82]. After the initial homing experiments, it was demonstrated that BM MNCs from non-chagasic syngeneic donors significantly reduced cardiac inflammation and fibrosis in mice with chronic *T. cruzi* infection. This improvement was maintained for up to 6 months after cell therapy [72, 83]. Using magnetic resonance imaging (MRI), it was demonstrated that BM-MNCs prevented and reversed the right ventricular dilatation induced by *T. cruzi* infection. Thus, histological improvement achieved by BM-MNC transplantation was paralleled by a functional correlate [72, 84].

One of the most striking observations after cell therapy in mice, chronically infected with *T. cruzi*, was related to the pattern of gene expression examined by microarray [72, 85]. While chagasic mice had 1702 out of 9390 (18%) cardiac genes with expression altered by infection after BM-MNC therapy, 96% of these genes were restored to normal levels although additional 109 genes had their expression altered by therapy [72, 85]. Transplantation of BM-MNCs in experimental mice showed immunomodulatory effects in chronic chagasic CMP and caused reversion of gene expression alterations in hearts of mice infected with *T. cruzi* [85]. In a chicken model of Chagas disease, the genetic alterations resulting from kDNA integration in the host genome caused an autoimmune-mediated destruction of heart tissues even in the absence of *T. cruzi* parasites [86]. In animal models of Chagas disease, microarrays were used to analyze global gene expression [87]. Eight distinct categories of mRNAs were found to be differentially regulated during infection and the dysregulation of several key genes was identified. These findings provide insights into the pathogenesis of chagasic CMP and provide new targets for intervention [87].

It is expected that within a few years, scientists will be able to find the (1) best animal model, (2) appropriate stem cell dose, (3) appropriate stem cell type, (4) injection route, and (5) disease status that will result in benefits for chronic chagasic CMP patients [72]. Despite the success of stem cell therapy in animals, preclinical models of Chagas disease have not translated into successful human studies [88]. Addressing the challenges associated with future research and providing solutions to have acceptable levels of safety and strict quality controls would enable successful clinical applications of stem cell therapies in chagasic CMP [88].

Ultimately, transplantation of BM-derived cells may prove to be an important therapeutic modality in the treatment of end-stage chagasic heart disease [89]. Identifying which cell type(s) is responsible for the effects observed in both animal studies and preliminary human experiments will be an important step in further improving this treatment [89]. Since the percentage of stem cells (either hematopoeitic or mesenchymal) is minimal in the mononuclear fraction, use of purified stem cell population has the potential to significantly increase the therapeutic benefit of cell therapy in chagasic CMP [89].

6. MSCS and fungal infections

Despite the availability of new antifungal agents, mortality and morbidity related to invasive fungal infections is still high, particularly in patients with hematological malignancies and in recipients of SOT and HSCT [90, 91]. The recent advances in the immunopathogenesis of invasive aspergillosis have provided critical information to augment host immunity against fungal pathogens [90, 91]. Potential approaches for enhancement of the host immune system to combat invasive fungal infections include (1) administration of effector and regulatory cells such as granulocytes, antigen-specific T-cells, natural killer cells and dendritic cells, and (2) administration of cytokines, interferons and growth factors such as INF-γ, G-CSF and granulocyte monocyte-colony stimulating factor (GM-CSF) [90, 91]. Although promising results have been obtained from animal studies, limited data are available to draw conclusions on risks and benefits in clinical settings [90, 91]. So, appropriately designed, multicenter and international clinical trials are needed in order to improve the outcome of invasive fungal infections in immunocompromised individuals [90, 91]. MSC therapy in HSCT recipients has been found to facilitate faster control of invasive fungal infections [92].

Studies have shown that human cathelicidin LL-37 and its fragments LL13-37 and LL 17-37 exhibit similar potencies in inhibiting the growth of *Candida albicans* (*C. albicans*) [93]. However, death of (*C. albicans*) cells may not solely be due to increased permeability caused by LL13-37, but can also be due to certain intracellular targets [93]. In patients with severe refractory neutrophilic bronchial asthma, administration of MSCs appears to play a significant role in decreasing inflammation and ameliorating disease manifestations by mediating *Aspergillus*-induced inflammation through inhibition of the Th_{17} signaling pathway [94].

7. MSCs and wound healing

Chronic non-healing wounds are a serious medical problem [95]. Biofilms, which are bacterial communities attached to a surface and become protected by a polysaccharide coating, are believed to contribute significantly to the persistence of chronic wounds by altering the host immune response. Current treatment options are ineffective and do not significantly target biofilms [95]. It has been shown that the paracrine factors secreted by reprogrammed MSCs accelerate wound healing and reduce bacterial growth in biofilm-infected wounds in mice [95]. Therefore, clinical studies are needed to test the efficacy of these paracrine factors in the eradication of biofilms in patients with non-healing wounds [95].

In an in vitro study, MSCs have recently been shown to exert measurable antimicrobial activities in a synovial fluid setting, suggesting that they may have future application as adjunct therapy for peri-prosthetic joint infections [96]. Also, AT-MSCs have been shown to

http://dx.doi.org/10.5772/60640

enhance closure of enterocutaneous fistula and aided in the recovery and healing of wounds in a rat model [97].

8. MSCs and lung injury

Melatonin, the chief secretary product of the pineal gland, is an indirect antioxidant that acts to stabilize cell membranes thereby making them less susceptible to oxidative insults and ultimately suppressing inflammatory reactions [98]. Combined melatonin and apoptotic AT-MSCs treatment has been found to be superior to either regimen alone in ameliorating lung injury in the setting of CLP-induced sepsis in a rodent model [98]. Also, adult tissue-derived MSCs have been shown to have antimicrobial effects that inhibit bacterial growth in lung tissues [96].

BM-derived MSCs have unlimited potential clinical applications in acute lung injury, due to various pulmonary disorders and they exert their effects by various mechanisms [99]. *Vibrio vulnificus* sepsis can induce acute lung damage and pulmonary edema. BM-derived MSCs can downregulate inflammatory cytokines and reduce lung injury caused by *Vibrio vulnificus* sepsis in mice [100]. BM-MNCs were able to diminish pulmonary inflammation, lung elastance, lung modeling and fibrosis resulting in lower mortality in acute respiratory distress syndrome (ARDS) experimental models [101]. However, the benefits of BM-MNCs depend on several factors including the type of initial insult as they were shown to exert different effects on endothelial cell activation and adhesion molecules. Therefore, further studies are needed to clarify their mechanisms and to examine this novel therapy in clinical trials [101].

Multipotent MSCs have shown remarkable therapeutic effects in preclinical models of both ARDS and sepsis [102]. Initial research focused on the ability of MSCs to engraft at sites of tissue injury [102]. Recent literature shows increasing evidence suggesting that MSCs exert their therapeutic effects through mechanisms that are unrelated to long-term incorporation into tissues of the host [102]. One of the most compelling pathways is their capacity to interact with injured tissue through the release of soluble bioactive factors [102].

9. MSCs and neutrophils

MSCs participate in the regulation of inflammation and innate immunity by responding to pathogen-derived signals and by regulating the function of innate immune cells [103]. MSCs derived from the BM and peripheral tissues share common basic cell-biological functions [103]. MSCs contribute to the resolution of infection and inflammation by promoting the antimicrobial activity of polymorphonuclear leukocytes, a property that is exhibited by MSCs derived from either BM or peripheral glandular tissues [103]. MSCs rescue neutrophils from nutrient- or serum-deprived cell death. Whether this effect is exerted through a specific signaling pathway or confining neutrophils in a resting state by MSCs requires further evaluation [33].

Both multipotent adult progenitor cells (MAPCs) and MSCs are adult stem cells that can be derived from BM and are currently utilized in tissue engineering due to their immunomodulatory and trophic effects [104]. The use of stem-cell-based immunotherapy is very promising as in vitro studies have shown comparable suppressive effects of both human MSCs and human MAPCs [104]. However, the broader expansion capacities of human MAPCs make them more attractive than human MSCs for clinical use [104]. Neutrophils release many lysis-inducing factors and cause local tissue damage. Conversely, neutrophils themselves could become a target in controlling sepsis [105].

10. Antimicrobial effects of MSCs

Multipotent, BM-MSCs (MP-BM-MSCs) are culture-expanded, nonhematopoietic stromal cells with immunomodulatory properties that are currently being investigated as novel cellular therapies in the prevention and treatment of clinical diseases that are associated with immune dysfunction [106]. Preclinical studies suggest that MP-BM-MSCs may protect against infectious challenge through direct or indirect pathways [106]. Direct effects exerted by MP-BM-MSCs on the pathogen are manifested by reduction of pathogen burden by inhibition of growth through soluble factors and enhancement of antimicrobial function of immune cells [106]. Indirect effects exerted by MP-BM-MSCs on the host include: (1) attenuation of proinflammatory cytokine and chemokine induction, (2) reduction of proinflammatory cells immigration into the sites of tissue injury or infection, and (3) reduction of immunoregulatory soluble and cellular factors that preserve organ function. These preclinical studies provide insights into the future therapeutic applications of MP-BM-MSCs [106].

Through toll-like receptor (TLR) signaling and immune crosstalk between MSCs and immune effector cells, MSCs become capable of maintaining a critical balance between (1) promotion of pathogen clearance during the initial phase of inflammatory response, and (2) suppression of inflammation to preserve host integrity and to facilitate tissue repair [106]. The presumed or suggested roles of MP-BM-MSCs during infection can be divided into five phases as shown in Table 4 [106].

Human BM-MSCs have direct antimicrobial activity, against Gram-positive as well as Gram-negative bacteria, which is mediated in part through the secretion of human cathelicidin hCAP-LL-37 [107]. Also, evidences suggest that MSCs position themselves between the acid fast bacilli harbored in the BM and the host protective T-cells. Hence MSCs can be a potential target for the treatment of latent TB [108]. Targeting MSCs is not expected to lead to the evolution of new resistance in the pathogen as MSCs do not need to be directly targeted and as the host immune response needs to be manipulated [108]. Upon stimulation of inflammatory cytokines, human MSCs exhibit broad-spectrum antimicrobial effector function directed against a range of clinically relevant bacteria, protozoal parasites and viruses [28]. Human MSCs act as immunosuppressants that concurrently exhibit potent antimicrobial effector function thus encouraging further evaluation in clinical trials [28]. Human multipotent MSCs exhibit broad-spectrum antimicrobial activity mediated by IDO [28].

Phase	Main effect / mechanism	Details and explanations
1	Detection of pathogens and damage signals	- Engagement of TLR-mediated signaling pathways - Recruitment of MSCs at the sites of infection
2	Activation of host immune response	- Maintenance of quiescent HSC pool - BM emigration of activated HSCs - Mobilization and emigration of immune effector cells from BM. - Thymic development to augment response of immune effector cells
3	Elimination of pathogens	- Production of microbiocidal soluble factors - Containment of infectious pathogen within micro-environment (pathogen phagocytosis)
4	Induction of proinflammatory gradients	- Antioxident soluble factor (HO-1) - Antiinflammatory soluble factors such as: IDO, PGFZ, IL-10, TGF-B, TGF-6, HLA-G5 and Galectin-1
5	Modulation of proinflammatory host immune response	- Activation of function, differentiation and migration of immune effector cells. - Augmentation of wound healing - Enhancement of revascularization - Inhibition of tissue toxicity - Modulation of inflammation and organ dysfunction.

TLR: toll-like receptor; HSC: hematopoietic stem cell; IL: interleukin; IDO: indoleamine-2, 4-dioxygenase

Table 4. Accepted roles MSCs during infection

Human gut microbial communities reside in an open ecosystem subject to disruptions ranging from dietary change to toxin exposure and pathogen invasion [109]. Host inflammatory mechanisms to remove harmful organisms and restrict bacteria to gut lumen commonly target conserved molecular patterns found on pathogens and commensals alike, yet healthy gut microbial communities can remain stable for years in humans [109]. A delicate balance between microbial resilience and host tolerance thus allows for commensal persistence throughout a diverse range of perturbations while preventing commensal overgrowth or depletion, either of which could have deleterious effects on the host [109]. Thus, antimicrobial peptide resistance mediates the resilience of prominent gut commensals during inflammation [109]. Stem cell differentiation can be regulated by TLR activation. Activation of TLRs on MSCs increases osteogenesis and decreases adipogenesis [110]. Upon exposure to microbial legends, stem cells change their differentiation to help the initial immune response [110]. Microbial ligands can influence stem cell fate via pattern recognition receptors. Microbial ligands such as LPS and lipoproteins can alter: proliferation, differentiation, migration and the function of stem cells. In mice stem cell proliferation is stimulated by TLR activation, while in humans most TLR activity does not seem to affect stem cell proliferation [110]. TLR activation alters stem cell differentiation, cytokine production of stem cells and immunosuppressive functions of MSCs

[110]. However, cytokine production can also be influenced by microenvironment, co-stimulatory molecules as well as downstream signaling pathways [110]. The immunosuppressive function of MSCs can be stimulated or inhibited by TLR activation depending on the type of signaling pathways that are activated [110]. With the identification of innate immune receptors on stem cells and the finding that microbial ligands can influence stem cell fate, a new era of stem cell research has begun [110].

11. MSCs and HSCT

Recent studies revealed that, during pathogen exposure, hematopoiesis may yield progeny in magnitudes that are different from those produced under routine or stable hemostatic circumstances [111]. HSCs not only sustain blood cell formation following bleeding, BM damage or chemotherapeutic ablation, but also respond directly to microbial products as well as inflammatory cytokines thus permitting real-time alterations in the direction of hematopoiesis in response to exposure to a certain burden of pathogenic organisms [111].

Various evidences from experimental models and clinical studies suggest the implication of bacterial and fungal infections in the pathogenesis of acute GVHD [112]. Appropriate treatment or prophylaxis of bacterial infections during the early post-HSCT period might be beneficial in reducing not only infection-related but also GVHD-related mortality [112]. To eliminate the systemic proinflammatory cytokine surge induced by bacterial infection, the following are needed: (1) specific antimicrobial strategies, (2) therapies targeting the pathways of innate immunity, and (3) nutritional interventions that might help in reducing the risk of acute GVHD. These issues should be evaluated prospectively in clinical trials[112]. In a retrospective cohort study that evaluated the risk factors for pneumonia-related death following HSCT and that included 691 patients, meta-analysis showed that the following were the risk factors that contributed to death: (1) acute GVHD grades II - IV, (2) CMV infection, and (3) receiving MSCs. Additionally, bacteremia and mold infections contributed to pneumonia-related deaths [113].

MSC therapy in HSCT recipients has been found to have the following advantages: (1) prevention and treatment of GVHD, (2) immune reconstitution, (3) induction of faster engraftment, (4) healing of infections and inflammatory disorders, and (5) reduction in the incidence of various infectious complications [92, 103, 104, 114, 115]. Studies have shown that administration of human BM-derived MSCs given during allogeneic bone marrow transplantation or human umbilical cord blood-derived MSCs in the post-transplant period, in patients with bone marrow failure, is safe and advantageous [92, 114]. In recipients of unrelated allogeneic HSCT following non-myeloablative conditioning therapy, co-infusion of MSCs did not show any pulmonary deterioration for up to 1 year post–transplant [115]. Additionally, in recipients of haploidentical HSCT, infusion of umbilical cord–derived MSCs did not increase the incidence of pulmonary infections [116].

12. Nanotechnology in stem cell research and therapy

Stem cell tracking using modern imaging modalities offers new insights into understanding the biology and achieving the full therapeutic potentials of stem cell therapies [117]. Currently used standard methods for tracking stem cells in vivo include: (1) MRI, (2) bioluminescence imaging, (3) positron emission tomography, (4) fluorescence imaging, (5) single-photon emission tomography, (6) X-ray computed microtomography, (7) Raman or surface enhanced Raman spectroscopy-based imaging, and (8) photoacoustic imaging [117-119]. The selection of the imaging modality and stem cell tracer or label should take into consideration: the clinical condition of the patient, the primary disease and the comorbid medical conditions [120].

Nanotechnology offers valuable information on migration, homing, survival, differentiation, function and the engraftment of transplanted stem cells [117, 121, 122]. The delivery of nanomaterials enables a personalized approach and individualized tailoring of nanotechnology and materials used to the specific needs of each patient [120]. To be of most use, tracking methods should ideally be non-invasive, high resolution and allow tracking in three dimensions [121]. To identify transplanted stem cells from the host tissue, optically active probes are usually used to label stem cells before their administration [117]. Nanofibers have recently gained substantial interest for their potential application in tissue engineering [123]. Nanofibers accommodate: (1) the survival and proliferation of human MSCs, and (2) the continuous differentiation of human MSCs into obsteoblasts and chondrocytes [123].

13. Homing and migration of MSCs

An important property of MSCs is their capacity for migration and homing in or around the zones damaged by trauma, inflammation, ischemia or tumor infiltration [124]. Intravenous administration of MSCs causes their migration, in substantial numbers, into areas of tissue damage [124]. Therefore, it is essential to understand the mechanisms involved in the homing and migration of MSCs. In addition, MSCs derived from various sources express different patterns of chemokines and their receptors [124]. Hence, the possibility to modulate the migration ability of MSCs opens a new era for the development of directed migration of greater numbers of MSCs injected intravenously into the sites of injury [124].

14. Banking of MSCs

MSCs hold great potential for developing effective cellular therapies and current trends indicate that their clinical applications will continue to rise markedly [125]. There has been considerable success in manufacturing and cryopreserving MSCs at laboratory level, but these successes have not translated into technologies developed at industrial scale [125]. The development of cost-effective and advanced technologies for the production and cryopreservation of MSCs is a crucial step in the process of successful clinical cell therapy [125]. Also, it is essential to have simple and appropriate but validated protocols for cell and tissue processing under good manufacturing conditions in order to develop a cost-effective banking of MSCs such as those obtained from umbilical cord blood [126].

15. Conclusions and future directions

MSCs are heterogeneous progenitor cells that have the capacity of self-renewal and multilineage differentiation. Their distinguished immunomodulation and immunosuppressive properties allowed MSCs to have several therapeutic applications. MSCs are essential constituents of the framework that supports organ integrity and tissue barriers. Suppression of both T and B cells enables them to be major players in controlling inflammatory response and in the innate response to bacterial infection. Human BM-MSCs possess direct antibacterial activity against Gram-negative bacilli and they have been shown to improve survival and reduce mortality in animal models having septic complications. Also, human BM-MSCs can act as drug delivery vehicles, enhance the effectiveness of conventional antimicrobials and may prevent the evolution of drug–resistant microbes. MSCs contain a subset of IL-17$^+$ that is capable of inhibiting the growth of *C. albicans*. CD 271$^+$ BM-MSCs may provide a long-term protective intracellular niche in the host where *M.TB* organisms remain viable but in a dormant state. Two recent clinical trials in humans have shown that autologous transplantation of MSCs can successfully treat MDR strains of *M.TB*. Animal studies have demonstrated that MSCs enhance host defenses against malaria. MSC therapy improves liver function and promotes hepatocellular regeneration in patients with hepatic fibrosis caused by schistosomiasis. Transplantation of MSCs has been shown to reverse right ventricular dilatation, cardiomyopathy and advanced heart involvement in *T. cruzi* infection. Transfusion of MSCs can confer resistance to HIV and may restore immune reconstitution in infected individuals. Autologous MSC transfusion in patients having liver cirrhosis secondary to hepatitis B or C infection improves liver function tests. MSCs improve murine models of acute myocarditis caused by CVB_3 infection. However, MSCs are not effective in the prevention or treatment of severe *Influenza virus* infections. There is low risk of transmission of human herpes viruses by transplantation of MSCs from healthy seropositive donors. *CMV* infection impairs the immunosuppressive and antimicrobial effector functions of human MSCs, thus overt *CMV* infection in recipients of HSCT may undermine the clinical efficacy of MSCs in treating GVHD. MSC therapy in patients with severe GVHD is associated with increased mortality related to *Adenovirus* infections. Therapeutic applications of BM-MSCs in recipients of HSCT include prevention and treatment of GVHD, induction of faster engraftment, immune reconstitution, healing of inflammation as well as treatment and prevention of various infectious complications. Also, cultured or banked human BM-MSCs may be effective in treating sepsis in high-risk patients.

Taking into consideration the remarkable success in the utilization of MSCs in the treatment of various infections in animal models and the few published clinical trials in humans with MDR infections, such as those caused by *M.TB*, and with the advancements in technology and medical care, it is reasonable to predict a similar success in the use of MSC therapy in humans in the near future. However, complications of this form of cellular therapy should never be underestimated. Also, it is essential to have banking facilities for these stem cells in addition to guidelines and protocols for their use in high risk individuals, particularly those with *HIV*, MDR-TB, hematological malignancy, recipients of SOT and HSCT as well as patients receiving advanced level of care in ICUs.

Author details

K.A. Al-Anazi[1*] and A.M. Al-Jasser[2]

*Address all correspondence to: kaa_alanazi@yahoo.com

1 Department of Adult Hematology and Hematopoietic Stem Cell Transplantation, Cancer Center, King Fahad Specialist Hospital, Dammam, Saudi Arabia

2 Riyadh Regional Laboratory, Ministry of Health, Riyadh, Saudi Arabia

References

[1] Wong RSY. Mesenchymal stem cells: angels or demons. J Biomed Biotechnol. 2011; 2011: 459510; 1–8.

[2] Ra JC, Shin IS, Kim SH, Kang SK, Kang BC, Lee HY, et al. Safety of intravenous infusion of human adipose tissue-derived mesenchymal stem cells in animals and humans. Stem Cells Dev. 2011; 20 (8) 1297–1308.

[3] Herberts CA, Kwa MS, Hermsen HP. Risk factors in the development of stem cell therapy. J Transl Med. 2011; 9 (29): 1–14.

[4] Wang Y, Han Z-b, Song Y-p, Han ZC. Safety of mesenchymal stem cells for clinical application. Stem Cell Int. 2012; 2012: 652034; 1–4.

[5] Lalu MM, McIntyre L, Pugliese C, Fergusson D, Winston BW, Marshall JC, et al. for the Canadian Trials Group. Safety of cell therapy with mesenchymal stromal cells (SafeCell): a systematic review and a meta-analysis of clinical trials. PLoS One 2012; 7 (10): e–47559.

[6] Cornėlio DA, de Medeiros SRB. Genetic evaluation of mesenchymal stem cells. Rev Bras Hematol Hemoter. 2014; 36(4): 238–240.

[7] von Bahr L, Sundberg B, Lönnies L, Sander B, Karbach H, Ljungman P, et al. Long-term complications, immuonological effects, and role of passage for outcome in mesenchymal stromal cell therapy. Biol Blood Marrow Transplant. 2012; 18 (4): 557–564.

[8] Sundin M, Lindblom A, Orvell C, Barrett AJ, Sundberg B, Watz E, et al. Persistence of human parvovirus B_{19} in multipotent mesenchymal stromal cells expressing the erythrocyte P antigen: implications for transplantation. Biol Blood Marrow Transplant. 2008; 14 (10): 1172–1179.

[9] Ning H, Yang F, Jiang M, Hu L, Feng K, Zhang J, et al. The correlation between cotransplantation of mesenchymal stem cells and higher recurrence rate in hematologic malignancy patients: outcome of a pilot clinical study. Leukemia. 2008; 22 (3): 593–599.

[10] Leentjens J, Kox M, van der Hoven JG, Netea MG, Dickkers P. Immunotherapy for the adjunctive treatment of sepsis: from immunosuppression to immunomodulation, time for a paradigm change? Am J Respir Crit Care Med. 2013; 187 (12): 1287–1293.

[11] Elman JS, Li M, Wang F, Gimble JM, Parekkadan B. A comparison of adipose and bone marrow-derived mesenchymal stromal cell secreted factors in the treatment of systemic inflammation. J Inflamm (Lond). 2014; 11(1):1.

[12] Shin S, Kim Y, Jeong S, Hong S, Kim I, Lee W; et al. The therapeutic effect of human adult stem cells derived from adipose tissue in endotoxemic rat model. Int J Med Sci. 2013; 10 (1): 8–18.

[13] Suen C, Cheung L. The world's first in-human stem cell trial for septic shock: a bench-to-bedside journey in critical care from the perspective of Dr Lauralyn McIntyre. UOJM 2013; 3: 8–10.

[14] Gorbunov MV, Elliott TB, McDaniel DP, Lund K, Liao P-J, Zhai M, et al. Up-regulation of autophagy defence mechanisms in mouse mesenchymal stromal cells in response to ionizing irradiation followed by bacterial challenge. Edited by Yannik Bailly; In–Tech; 2013; ISBN: 978-953-51-1062-0.

[15] Mai S, Khan M, Liaw P, Fox-Robichaud A, on behalf of the Canadadian Critical Care Translational Biology Group (CCCTBG). Experimental Sepsis Models. Edited by Luciano Azevedo; In–Tech; 2012. ISBN: 978-953-51-0780-4.

[16] Azevedo LCP. Microparticles and exosomes: are they part of important pathways in sepsis pathophysiology? Severe sepsis and septic shock-understanding a serious killer. Edited by Dr Richardo Fernandez; In–Tech; 2012; ISBN: 978-953-307-950-9.

[17] Meisel R, Heseler K, Nau J, Schmidt SK, Leinewever M, Pudelko S, et al. Cytomegalovirus infection impairs immunosuppressive and antimicrobial effector functions of human multipotent mesenchymal stromal cells. Mediators Inflamm. 2014; 2014: 898630.

[18] Van Linthout S, Savvatis K, Miteva K, Peng T, Ringe J, Warstat K, et al. Mesenchymal stem cells improve murine acute *Coxsackivirus* B_3-induced myocarditis. Eur Heart J. 2011; 32: 2168–2178.

[19] Hall SR, Tsoyi K, Ith B, Padera RF Jr, Leaderer JA, Wang Z, et al. Mesenchymal stromal cells improve survival during sepsis in the absence of neutrophils. Stem Cells. 2013; 31 (2): 397-407.

[20] Mei SHJ, Haitsma JJ, Dos Santos CC, Deng Y, Lai PFH, Slutsky AS, et al. Mesenchymal stem cells reduce inflammation while enhancing bacterial clearance and improving survival in sepsis. Am J Respir Crit Care Med. 2010; 182 (8):1047–1057.

[21] Luo C-J, Zhang F-j, Zhang L, Geng Y-q, Li Q-g, Hong Q, et al. Mesenchymal stem cells ameliorate sepsis-associated acute kidney injury in mice. Shock. 2014; 41 (2), 123–129.

[22] Gupta N, Krasnodembskaya A, Kapetanaki M, Mouded M, Tan X, Serikov V, et al. Mesenchymal stem cells enhance survival and bacterial clearance in murine *Escherichia coli* pneumonia. Thorax 2012; 67 (6): 533–539.

[23] Nemeth K, Leelahavanichkul A, Yuen PST, Mayer B, Parmelee A, Doi K, et al. Bone marrow stromal cells attenuate sepsis via prostaglandin E_2- dependent reprogramming of host macrophages to increase their interleukin-10 production. Nat Med. 2009; 15 (1): 42–49.

[24] Gonzalez-Ray E, Gonzalez MA, Rico L, Buscher D, Delgado M. Human adult stem cells derived from adipose tissue protect against experimental colitis and sepsis. Gut 2009; 58 (7): 929–939.

[25] Chang C-L, Leu S, Sung H-C, Zhen Y-Y, Cho C-L, Chen A, et al. Impact of apoptotic adipose-derived mesenchymal stem cells on attenuating organ damage and reducing mortality in rat sepsis syndrome induced by cecal puncture and ligation. J Transl Med. 2012; 10: 244.

[26] Yang R, Liu Y, Kelk P, Qu C, Akiyama K, Chen C, et al. A subset of IL-17^+ mesenchymal stem cells possesses anti-*Candida albicans* effect. Cell Res 2013; 23 (1):107–121.

[27] Sung P-H, Chang C-L, Tsai T-H, Chang L-T, Leu S, Chen Y-L, et al. Apoptotic adipose-derived mesenchymal stem cell therapy protects against lung and kidney injury in sepsis syndrome caused by cecal ligation puncture in rats. Stem Cell Res Ther. 2013; 4 (6): 155.

[28] Meisel R, Brookers S, Heseler K, Degistirici Ö; Bülle H, Woite C, et al. Human but not murine multipotent mesenchymal stromal cells exhibit broad-spectrum antimicrobial effector function mediated by indoleamine 2, 3-dioxygenase. Leukemia 2011; 25: 648–654.

[29] Krasnodesmbskaya A, Song Y, Fang X, Gupta N, Serikov V, Lee J-W, et al. Antibacterial effect of human mesenchymal stem cells is mediated in part from secretion of the antimicrobial peptide LL-37. Stem Cells 2010; 28 (12): 2229–2238.

[30] Krasnodesmbskaya A, Samarani G, Song Y, Zhuo H, Su X, Lee J-W, et al. Human mesenchymal stem cells reduce mortality and bacteremia in Gram-negative sepsis in

mice in part by enhancing the phagocytic activity of blood monocytes. Am J Physiol Lung Cell Mol Physiol. 2012; 302 (10): L1003–L1013.

[31] Kim ES, Chang YS, Choi SJ, Kim TK, Yoo HS, Ahn SY, et al. Intratracheal transplantation of human umbilical cord blood-derived mesenchymal stem cells attenuates *Escherichia coli*-induced acute lung injury in mice. Respir Res. 2011; 12 (108): 1-11.

[32] Cassatella MA, Mosna F, Micheletti A, Lisi V, Tamassia N, Cont C, et al. Toll-like receptor-3-activated human mesenchymal stromal cells significantly prolong the survival and function of neutrophils. Stem Cells 2011; 29: 1001–1011.

[33] Maqbool M, Vidyadaran S, George E, Ramasamy R. Human mesenchymal stem cells protect neutrophils from serum-deprived cell death. Cell Biol Int. 2011; 35: 1247–1251.

[34] Raffaghello L, Bianchi G, Bertolotto M, Montecucco F, Busca A, Dallegri F, et al.. Human mesenchuymal stem cells inhibit neutrophil apoptosis: a model for neutrophil preseveation in the bone marrow niche. Stem Cells 2008; 26: 151–162.

[35] Lee JW, Krasnodembskaya A, McKenna DH, Song Y, Abbott J, Matthay MA. Therapeutic effects of human mesenchymal stem cells in ex vivo human lungs injured with live bacteria. Am J Respir Crit Care Med. 2013; 187 (7): 751–760.

[36] Kusadasi N, Groeneveld AB. A perspective on mesenchymal stromal cell transplantation in the treatment of sepsis. Shock 2013; 40 (5): 352–357.

[37] Yuan Y, Lin S, Guo N, Zhao C, Shen S, Bu X, et al. Marrow mesenchymal stromal cells reduce methicillin-resistant *Staphylococcus aureus* infection in rat models. Cytotherapy 2014; 16 (1): 56–63.

[38] Kim H, Darwish I, Monroy M-F, Prockop DJ, Liles WC, Kain KC. Mesenchymal stromal (stem) cells suppress pro-inflammatory cytokine production but fail to improve survival in experimental *Staphylococcal* toxic shock syndrome. BMC Immunol. 2014; 15:1.

[39] Johnson V, Webb T, Dow S. Activated mesenchymal stem cells amplify antibiotic activity against chronic *Staphylococcus aureus* infection (P2056). J Immunol. 2013; 190: 180, 11.

[40] Dow S, Johnson V. Activated mesenchymal stem cells enhance antibiotic treatment of bacterial infections. Colorado State University Ventures, Inc. Posted by Sarah Belford on 14/11/2013; file No.: 12-060.

[41] Weil BR, Herrmann JL, Abarbancell AM, Manukyan MC, Poynter JA, Meldrum DR. Intravenous infusion of mesenchymal stem cells is associated with improved myocardial function during endotoxemia. Shock 2011; 36 (3): 235–241.

[42] Sisto F, Bonomi A, Cavicchini L, Cocce V, Scaltrito MM, Bondiolotti G, et al. Human mesenchymal stromal cells can uptake and release ciprofloxacin, acquiring in vitro anti-bacterial activity. Cytotherapy 2014; 16 (2): 181–190.

[43] Todd GP, LeRoux MA, Danilkovitch-Miagkova A. Mesenchymal stem cells as vehicles for targeted therapies. Drug discovery and development-present and future. Edited by Dr Izet Kapetanovic; In–Tech; 2011. ISBN: 978-953-307-615-7.

[44] Avanzi S, Leoni V, Rotola A, Alviano F, Solimando L, Lanzoni G, et al. Susceptibility of human placenta derived mesenchymal stromal/stem cells to human herpesviruses infection. PLoS One. 2013; 8 (8): e–71412.

[45] Wei G, Lin M, Cai Z, Huang H. Cytomegalovirus infection in mesenchymal stem cells and their activation could be enhanced by nuclear factor-kB inhibitor pyrrolidindithiocarbamate in vitro. Transplant Proc. 2011; 43 (5): 1944–1949.

[46] Smirnov SV, Harbacheuski R, Lewis-Antes A, Zhu H, Rameshwar P, Kotenko SV. Bone marrow-derived mesenchymal stem cells as a target for cytomegalovirus infection: implications for hematopoiesis, self-renewal and differentiation potential. Virology 2007; 360 (1): 6–16.

[47] Cotter EJ, Maughan RT, Doran PP. Role of mesenchymal stem cells (MSC) in HIV-1 associated bone and lipid toxicities. Stem cells and cancer stem cells. Edited by M.A. Hayat; Springer; 2012; 8: 79–90.

[48] Zhang Z, Fu J, Xu X, Wang S, Xu R, Zhao M. Safety and immunological responses to human mesenchymal stem cell therapy in difficult-to-treat HIV-1-infected patients. AIDS. 2013; 27 (8): 1283–1293.

[49] Allam O, Samarani S, Ahmad A. Mesenchymal stem cell therapy in HIV-infected HAART-treated nonimmune responders restores immune competence. AIDS. 2013; 27 (8): 1349–1352.

[50] Rodrigues ES, do Carmo Favarin M, de Macedo MD, Otaguiri KK, Orellana MD, Takayanagui OM. HTLV-1 infection in bone marrow mesenchymal stem cells isolated from HTLV-1 individuals. Retrovirology. 2014; 11 (Suppl l): P106.

[51] Darwish I, Banner D, Mubareka S, Kim H, Besla R, Kelvin DJ, et al. Mesenchymal stromal (stem) cell therapy fails to improve outcomes in experimental severe influenza. PLoS One. 2013; 8: e–71761.

[52] Zhao C, Wu N, Deng F, Zhang H, Wang N, Zhang W. *Adenovirus*-medicated gene transfer in mesenchymal stem cells can be significantly enhanced by the cationic polymer polybrene. PLoS One. 2014; 9 (3): e–92908.

[53] Ma R, Xing Q, Shao L, Wang D, Hao Q, Li X. Hepatitis B virus infection and replication in human bone marrow mesenchymal stem cells. Virology J. 2011; 8 (486): 1–8.

[54] Zhang Y. Persisters, persistent infections and Yin-Yang model. Emerg Microbes Infect. 2014; 3 (1): e–3.

[55] Sikri K, Tyagi JS. The evolution of *Mycobacterium tuberculosis* dormacy models. Current Science. 2013; 105 (5): 607–616.

[56] Raghuvanshi S, Sharma P, Singh S, Kaer LV, Das G. *Mycobacterium tuberculosis* evades host immunity by recruiting mesenchymal stem cells. Proc Natl Acad Sci U S A. 2010; 107 (50): 21653–21658.

[57] Al-Anazi KA, Al-Jasser AM, Alsaleh K. Infections caused by *Mycobacterium tuberculosis* in recipients of hematopoietic stem cell transplantation. Front Oncol. 2014; 4: 231.

[58] Kelly DM, ten Bokum AMC, O'Leary SM, O'Sullivan MP, Keane J. Bystander macrophage apoptosis after *Mycobacterium tuberculosis* H37 Ra infection. Infect Immun. 2008; 76 (1): 351–360.

[59] Tousif S, Singh DK, Ahmad S, Moodley P, Bhattacharyya M, Van Kaer L, et al. Isoniazid induces apoptosis of activated $CD4^+$ T cells: implications for post-therapy tuberculosis reactivation and reinfection. J Biol Chem. 2014; 289 (44): 30190–30195.

[60] Das B, Kashino SS, Pulu I, Kalita D, Swamy V, Yeger H, et al. $CD271^+$ bone marrow mesenchymal stem cells may provide a niche for dormant *Mycobacterium tuberculosis*. Sci Transl Med. 2013; 5 (170): 170 ra13.

[61] Beamer G, Major S, Das B, Campos-Neto A. Bone marrow mesenchymal stem cells provide an antibiotic-protective niche for persistent viable *Mycobacterium tuberculosis* that survive antibiotic treatment. Am J Pathol. 2014; 184 (12): 3170–3175.

[62] Erokhin VV, Vasil'eva IA, Konopoliannikov AG, Chukanov VI, Tsyb AF, Bogdasarian TR. Systemic transplantation of autologous mesenchymal stem cells of the bone marrow in the treatment of patients with multidrug-resistant pulmonary tuberculosis. Probl Tuberk Bolezn Leak. 2008; 10: 3–6.

[63] Skrahin A, Ahmed RK, Ferrata G, Rane L, Poiret T, Isaikina Y, et al. Autologous mesenchymal stromal cell infusion as part of adjunct treatment in patients with multidrug and extensively drug-resistant tuberculosis: an open-label phase 1 safety trial. Lancet Respir Med 2014; 2 (2): 108–122.

[64] Thakur RS, Tousif S, Awasthi V, Sanyal A, Atul PK, Punia P, et al. Mesenchymal stem cells play an important role in host protective immune responses against malaria by modulating regulatory T cells. Eur J Immunol. 2013; 43: 2070–2077.

[65] Sinha S, Medhi B, Sehgal R. Challenges of drug-resistant malaria. Parasite 2014; 21(61): 1-15.

[66] Allahverdiyev A, Bagirova M, Elcicek S, Koc RC, Baydar SY, Findikli N, et al. Adipose tissue-derived mesenchymal stem cells as a new host cell in latent leishmaniasis. Am J Trop Med Hyg. 2011; 85 (3): 535–539.

[67] Bodgan C, Donhauser N, Döring R, Röllinghoff M, Diefenbach A, Ritting MG. Fibroblasts as host cells in latent leishmaniasis. J Exp Med. 2000; 191 (12): 2121–2129.

[68] Xu H, Qian H, Zhu W, Zhang X, Yan Y, Mao F. et al. Mesenchymal stem cells relieve fibrosis of *Schistosoma japonicum*-induced mouse liver injury. Exp Biol Med. 2012; 237 (5): 585–592.

[69] Zhang Y, Mi JY, Rui YJ, Xu YL, Wang W. Stem cell therapy for the treatment of parasitic infections: is it far away? Parasitol Res. 2014; 113 (2): 607–612.

[70] Abdul Aziz MT, Atta HM, Roshdy NK, Rashed LA, Sabry D, Hassouna AA, et al. Amelioration of murine *Schistosoma mansoni* induced liver fibrosis by mesenchymal stem cells. J Stem Cell Reg Med. 2012; 8 (1): 28–34.

[71] Spekker K, Leineweber M. Antimicrobial effects of murine mesenchymal stromal cells directed against *Toxoplasma gondii* and *Neospora caninum*: role of immunity-related GTPases (IRGs) and guanylate-binding proteins (GBPs). Med Microbiol Immunol. 2013; 202 (3): 197–206.

[72] de Carvalho AC, Carvalho AB, Mello DB, Goldenberg RC. Bone marrow-derived cell therapy in chagasic cardiac disease: a review of pre-clinical and clinical results. Cardiovasc Diagn Ther. 2012; 2 (3): 213–219.

[73] Muratore CA, Baranchuk A. Current and emerging therapeutic options for the treatment of chronic chagasic cardiomyopathy. Vasc Health Risk Manag 2010; 6: 593–601.

[74] Makino S, Fukuda K, Miyoshi S, Konishi F, Kodama H, Pan J, et al. Cardiomyocytes can be generated from bone marrow stromal cells in vitro. J Clin Invest. 1999; 103 (5): 697–705.

[75] Tomita, S, Li R-K, Weisel RD, Mickle DA, Kim E-J, Sakai T, et al. Autologous transplantation of bone marrow cells improves damaged heart function. Circulation 1999; 100 (Suppl II): II-247–II-256.

[76] Orlic D, Kajstura J, Chimenti S, Jakonluk I, Anderson SM, Li B, Plckel J, et al. Bone Marrow cells regenerate infarcted myocardium. Nature 2001; 410: 701–705.

[77] Silva JD, Souza MC, Antunes MA, Xisto DG, Padua TN, Abreu TP, et al. Effects of bone marrow derived mesenchymal stromal cells on lung and distal organ damage in experimental malaria. Am J Respir Crit Care Med. 2014; 189(1): A5249.

[78] Guarita-Souza LC, Carvalho KA, Woituwicz V, Rebelatto C, Senegaglia A, Hansen P, et al. Simultaneous autologous transplantation of cocultured mesenchymal stem cells and skeletal myoblasts improves ventricular function in a murine model of Chagas disease. Circulation 2006; 114 (Suppl I): I-120–I-124.

[79] Larocca TF, de Freitas Souza BS, Silva CA, Kaneto CM, de Alcantara AC, Azevedo C-M, et al. Transplantation of adipose-derived stem cells in experimental chronic chagasic cardiopathy. Arq Bras Cardiol. 2013; 100 (5): 460–468.

[80] Vilas-Boas F, Feitosa GS, Soares MB, Mota A, Pinho-Filho JA, Almeida AJ, et al. Early results of bone marrow cell transplantation to the myocardium of patients with heart failure due to Chagas disease. Arq Bras Cardiol. 2006; 87 (2): 1–7.

[81] Macambira SG, Vasconcelos JF, Costa CR, Klein W, Lima RS, Guimaräes P, et al. Granulocyte colony-stimulating factor treatment in chronic Chagas disease: preservation and improvement of cardiac structure and function. FASEB J. 2009; 23 (11): 3843–3850.

[82] Ribeiro Dos Santos RR, Rassi S, Feitosa G, Grecco OT, Rassi A Jr, da Cunha A, et al. Cell therapy in Chagas cardiomyopathy (Chagas arm of the multicenter randomized trial of cell therapy in cardiopathies study): a multicenter randomized trial. Circulation 2012; 125 (20): 2454–2461.

[83] Soares MB, Lima RS, Rocha LL, Takyia CM, Pontes-de-Carvalho L, de Carvalho AC, et al. Transplanted bone marrow cells repair heart tissue and reduce myocarditis in chronic chagasic mice. Am J pathol. 2004; 164 (2): 441–447.

[84] Goldenberg RC, Jelicks LA, Fortes FS, Weis LM, Rocha LL, Zhao D, et al. Bone marrow cell therapy ameliorates and reverses chagasic cardiomyopathy in a mouse model. J Infect Dis. 2008; 197 (4): 544–547.

[85] Soares MB, Lima RS, Souza BS, Vasconcelos JF, Rocha LL, Dos Santos RR, et al. Reversion of gene expression alterations in hearts of mice with chronic chagasic cardiomyopathy after transplantation of bone marrow cells. Cell Cycle 2011; 10 (9): 1448–1455.

[86] Teixeira AR, Gomes C, Nitz N, Sousa AO, Alves RM, Guimaro MC, et al. *Trypanosoma cruzi* in the chicken model: chagas-like heart disease in the absence of parasitism. PLoS One Negl Trop Dis. 2011; 5 (3): e–1000.

[87] Mukherjee S, Nagajyothi F, Mukhopadhyay A, Machado FS, Belbin TJ, , de Carvalho AC, et al. Alterations in myocardial gene expression associated with experimental *Trypanosoma cruzi* infection. Genomics 2008; 91: 423–432.

[88] de Calvalho KA, Abdelwahid E, Ferreira RJ, Irioda AC, Guarita-Souza LC. Preclinical stem cell therapy in Chagas disease: perspectives for future research. World J Transpl. 2013; 3 (4): 119–126.

[89] de Carvalho AC, Goldenberg RC, Jelicks LA, Soares MB, Dos Santos RR, Spray DC, et al. Cell therapy in Chagas disease. Interdiscip Perspect Infect Dis. 2009; 2009: 484358.

[90] Lehrnbecher T, Kalkum M, Champer J, Tramsen L, Schmidt S, Klingebiel T. Immunotherapy in invasive fungal infection–focus on invasive aspergillosis. Curr Pharm Des. 2013; 19 (20): 3689–3712.

[91] Lehrnbecher T, Tramsen L, Koehl U, Schmidt S, Bochennek K, Klingbiel T. Immunotherapy against invasive fungal diseases in stem cell transplant recipients. Immunol Invest. 2011; 40 (7-8): 839–852.

[92] Özdoğul H, Yeral M, Boğa C, Kozanoğlu I. Use of mesenchymal cells to modulate immune suppression and immune reconstruction in a patient with aplastic anemia complicated by invasive sino-orbital aspergillosis. Turk J Hematol. 2014; 31: 181–183.

[93] Wong JH, Ng TB, Legowska A, Rolka K, Hui M, Cho CH. Antifungal action of human cathelicidin fragment (LL13-37) on *Candida albicans*. Peptides 2011; 32 (10): 1996–2002.

[94] Lathrop MJ, Brooks EM, Bonenfant NR, Sokocevic D, Borg ZD, Goodwin M, et al. Mesenchymal stromal cells mediate *Aspergillus* hyphal extract-induced allergic airway inflammation by inhibition of the Th_{17} signaling pathway. Stem Cells Transl Med. 2014; 3: 194–205.

[95] Smith AA, Bellows CF. The use of paracrine factors from reprogrammed mesenchymal stem cells to treat biofilm-infected wound in vivo. Basic Science, Clinical Congress. October 2013, Washington, DC.

[96] Chen AF, Tuan RS. In vitro testing of adult mesenchymal stem cells as an adjunct therapy for treating periposthetic joint infections. Poster No. 0918. ORS 2012 Annual Meeting.

[97] Volpe BB, Santos Duarte Ads, Ribeiro TB, Atocchero I, Kharmandayan P, Olalla Saad ST, et al. Mesenchymal stromal cells from adipose tissue attached to suture material enhance the closure of enterocutaneous fistulas in a rat model. Cytotherapy 2014; 16 (12): 1709–1719.

[98] Chen H-H, Chang, C-L, Lin K-C, Sung P-S, Chaig H-T, Zhen Y-Y. Melatonin augments apoptotic adipose-derived mesenchymal stem cell treatment against sepsis-induced acute lung injury. Am J Transl Res. 2014; 6 (5): 439–458.

[99] Xia P, Peng X. Study on MSCs transplantation in respiratory diseases. J Chem Pham Res. 2014; 6 (4): 447–449.

[100] Chen X, Liang H, Lian J, Lu Y, Li X, Zhi S. The protective effect of bone marrow mesenchymal stem cell on lung injury induced by *Vibrio vulnificus* sepsis. Zhonghua Wei Zhong Bing Ji Jiu Yi Xue. 2014; 26 (11): 821–826.

[101] Pelosi P, Suthersan Y. Bone marrow-derived monocuclear cell therapy in sepsis-induced acute respiratory distress syndrome: different insults, different effects! Stem Cell Res Ther. 2013; 4 (143): 1–3.

[102] Walter J, Ware LB, Matthay MA. Mesenchymal stem cells: mechanisms of potential therapeutic benefit in ARDS and sepsis. Lancet respir Med. 2014; 2 (12): 1016–1026.

[103] Brandau S, Jakob M, Brukere KK, Bootz F, Giebel B, Radtke S. Mesenchymal stem cells augment the anti-bacterial activity of neutrophil granulocytes. PLoS One. 2014; 9 (9): e–106903.

[104] Jacobs SA, Roobrouck VD, Verfaillie CM, Van Gool SV. Immunological characteristics of human mesenchymal stem cells and multipotent adult progenitor cells. Immunol Cell Biol. 2013; 91: 32–39.

[105] Brown KA, Treacher DF. Neutrophils as potential therapeutic targets in sepsis. Discov Med. 2009; 6 (33): 118–122.

[106] Auletta JJ, Deans RJ, Bartholomew AM. Emerging roles of multipotent, bone marrow-derived stromal cells in host defense. Blood 2012; 119 (8): 1801–1809.

[107] Krasnodembskaya A, Song Y, Lee J-W, Matthay MA. Human mesenchymal stem cells exert antimicrobial activity in vitro and in vivo in part through the secretion of the antimicrobial peptide LL-37. Meeting Abstract: 183, 1: A 1246, 2011.

[108] Rai RC, Battacharya D, Das G. Stem cells in infectious diseases. Insight and control of infectious disease in global scenario. Edited by Dr Roy Priti; In–Tech 2012; ISBN: 978-953-51-0319-6.

[109] Cullen TW, Schofield WB, Barry NA, Putnam EF, Rundell EA, Trent MS, et al. Antimicrobial peptide resistance mediates resilience of prominent gut commensals during inflammation. Science 2015; 347: 170–175.

[110] Bouwman LI. Microbial ligands alter the fate of stem cells. TLR activation influences survival, proliferation, differentiation and function of stem cells. Universiteit Ulrecht, 2009; 1–28.

[111] Boika JR, Borghesi L. Hematopoiesis sculpted by pathogens: Toll-like receptors and inflammatory mediators directly activate stem cells. Cytokine. 2012; 57: 1–8.

[112] Fuji S, Kapp M, Einsele H. Possible implications of bacterial infections in acute graft-versus-host disease after allogeneic hematopoietic stem cell transplantation. Front Oncol. 2014; 4 (89): 1–7.

[113] Forslow U, Blennow O, Le Blanc K, Ringden, O, Gustafsson T, Mattson T, et al. Treatment with mesenchymal stromal cells is a risk factor for pneumonia-related death after allogeneic hematopoietic stem cell transplantation. Eur J Haematol. 2012; 89 (3): 220–227.

[114] Si Y, Yang K, Qin M, Zhang C, Du Z, Zhang X, et al. Efficacy and safety of human umbilical cord derived mesenchymal stem cell therapy in children with severe aplastic anemia following allogeneic hematopoietic stem cell transplantation: a retrospective case series of 37 patients. Pediatr Hematol Oncol. 2014; 31(1): 39–49.

[115] Moermans C, Lechanteur C, Boudoux E, Giet O, Henket M, Seidel L, et al. Impact of cotransplantation of mesenchymal stem cells on lung function after unrelated alloge-

neic hematopoietic stem cell transplantation following non-myeloablative conditioning. Transplantation 2014; 98 (3): 348–353.

[116] Han DM, Wang ZD, Ding L, Zheng XL, Yan HM, Xue M. Effect of umbilical cord MSC infusion on the pulmonary infection in haploidentical hematopoietic stem cell transplantation. Zhongguo Shi Yan Xue Ye Xue Xa Zhi 2014; 22 (4): 1084–1088.

[117] Gao Y, Cui Y, Chan JK, Xu C. Stem cell tracking with optically active nanoparticles. Am J Nucl Med Imaging. 2013; 3 (3): 232–246.

[118] Li SC, Tachili LM, Luo J, Dethlefs DA, Chen Z, Loudon WG. A biological global positioning system: considerations for tracking stem cell behaviors in the whole body. Stem Cell Rev. 2010; 6 (2): 317–333.

[119] Villa C, Erratico S, Razini P, Fiori F, Rustichelli F, Torrente Y, et al. Stem cell tracking by nanotechnologies. Int J Mol Sci. 2010; 11: 1070–1081.

[120] Janowski M, Bulte JW, Walczak P. Personalized nanomedicine advancements for stem cell tracking. Adv Drug Deliv Rev. 2012; 64 (13): 1488-1507.

[121] Edmundson M, Thanh NT, Song B. Nanoparticles based stem cell tracking in regenerative medicine. Theranostics 2013; 3 (8): 573–582.

[122] Wang Y, Xu C, Ow H. Commercial nanoparticles for stem cell labelling and tracking. Theranostics 2013; 3 (8): 544–560.

[123] Xin X, Hussain M, Mao JJ. Continuing differentiation of human mesenchymal stem cells and induced chondrogenic and osteogenic lineages in electrospun PLGA nanofiber scaffold. Biomaterials 2007; 28 (2): 316–325.

[124] Kholodenko IV, Konieva AA, Kholodenko RV, Yarygin KN. Molecular mechanisms of migration and homing of intravenously transplanted mesenchymal stem cells. J Regen Med Tissue Eng 2013; 2 (4): 1–11.

[125] Thirumala S, Goebel WS, Woods EJ. Manufacturing and banking of mesenchymal stem cells. Expert Opin Biol Ther. 2013; 13 (5): 673–691.

[126] Cooper K, Viswanathan C. Establishment of a mesenchymal stem cell bank. Stem Cells Int. 2011; 2011: 905621.